WITHDRAWN

L.V. Kantorovich:
Essays in
Optimal Planning

L.V. Kantorovich: Essays in Optimal Planning

Selected with an Introduction by Leon Smolinski

INTERNATIONAL ARTS AND SCIENCES PRESS, INC., WHITE PLAINS, N.Y.

Translated by Arlo Schultz and others.

Published simultaneously as Volume XIX, No. 4-6 of <u>Problems
of Economics</u>, edited by Murray Yanowitch, Leonard J. Kirsch,
and Elena Kirsch.

Essays 1, 2, 3, and 14 are published by arrangement with
VAAP, the USSR Copyright Agency.

Library of Congress Catalog Card Number: 75-46110.
International Standard Book Number: 0-87332-076-X.

Printed in the United States of America.

Contents

77- 2966

L. V. Kantorovich and Optimal Planning

I

A young professor of mathematics at Leningrad University
was approached by the local plywood trust in 1938 to assist in
the solution of a seemingly trivial but puzzling problem in production scheduling. How to draw a work schedule for eight
lathes of various output capacities so as to maximize the total
output of five varieties of plywood, subject to the constraint of
a given product mix? The trust's laboratory was unable to arrive at a satisfactory solution that could not be further improved
upon. Could Professor Kantorovich tell them where they had
gone wrong?

Born in 1912, Leonid Vital'evich Kantorovich was only twenty-
six years old when the plywood trust consulted him, but he had
already acquired an international reputation for his creative
work in pure and applied mathematics. He had enrolled in the
mathematics department of Leningrad University at the age of
fourteen, graduated at eighteen, become a full professor at
twenty-two, and received the rare and coveted degree of doctor
of science in 1935. (1) Faced with "the plywood trust problem,"
as it is now known in the history of mathematical economics,
he promptly recognized it as belonging to a wide class of extremal problems with linear constraints which are extremely
common in resource allocation in a planned economy. In effect,
is not maximization under constraints the essence of economic
planning? The mathematical formulation of such problems is
simple, but their solution by means of conventional mathematical
analysis utilizing Lagrange multipliers is "utterly inapplicable

in practice, since it requires the solution of tens of thousands if not millions of systems of simultaneous equations." (2)

A run-of-the-mill mathematician would have terminated his consultation then and there. But Kantorovich took up the challenge and devised a new method for solving such problems by means of what he called "solving multipliers" (or, in Western terminology, shadow prices). This discovery, presented in his 1939 monograph "The Mathematical Methods of Organizing and Planning Production" [1], gave rise to a new branch of mathematical economics and of applied mathematics. After its independent discovery and development in the USA by George Bernard Dantzig in 1947, this new discipline came to be known as linear programming, a term suggested by T. J. Koopmans, with whom Kantorovich was to share the Nobel Prize in economics decades later.

Even a mathematical genius might have been content with this result, which led Kantorovich to the solution of the celebrated eighteenth-century Monge's Problem (3), no mean achievement in itself. But Kantorovich, with an intuition rare in a mathematician who had never been exposed to the economic theory of resource allocation, saw immediately the broad applicability of his solving multipliers to optimal choices in a wide variety of economic situations. Koopmans views this insight as a great accomplishment in its own right: "The wide range of applications perceived by the author makes his paper an early classic in the science of management under any economic system." He concludes: "The paper stands as a highly original contribution of the mathematical mind to problems which few at that time would have perceived as mathematical in nature — on a par with the earlier work of von Neumann on proportional economic growth in a competitive market economy and the later work of G. B. Dantzig." (4)

In linear programming the optimal plan (be it for a factory, an industry, or a nation) is seen as a solution of a system of linear forms with linear constraints.

The purpose of optimal planning is to maximize the value of

the objective function given the resource vector and the technology matrix. In keeping with the philosophy of Soviet economic planning, the objective function is defined as maximizing total value of output of n products in given proportions

$$k_1, k_2, \ldots, k_n.$$

Let a technology s being applied at a unit level be described by a vector

$$a^s = (a_1^s, \ldots, a_n^s, a_{n+1}^s, \ldots, a_N^s), s = 1, 2, \ldots, r$$

whose first n components are final products and last $N - n$ components are resources.

To construct a plan means to find a vector $\pi = (x_1, \ldots, x_r)$ each of whose nonnegative components refers to the intensity of utilization of one of the available technologies.

Output of the i-th commodity can be then described as:

$$y_i^\pi = \sum_{s=1}^r a_i^s x_s, i = 1, 2, \ldots, n.$$

The necessary and sufficient condition for the plan π to be feasible is that all technologies should be used with nonnegative intensities and that the amounts of resources required to operate them should not exceed the amounts actually available.

A feasible plan π is optimal if total output produced is at a maximum, viz:

$$\mathfrak{M}(\pi) = \min_{1 \le i \le n} \frac{y_i}{k_i}, i = 1, 2, \ldots, n$$

The necessary and sufficient condition for an optimal plan is the existence of a system of shadow prices, found as a solution of the dual of the linear programming problem, which satisfies the following conditions:

1) shadow prices are nonnegative, and the shadow price for at least one final product included in the optimal plan must be positive;

2) for each technology the total value of final products does not exceed their total resource cost, both being calculated in terms of their shadow prices;

3) for each technology actually adopted in the optimal plan, total value of outputs equals the sum total of factor returns, resulting in zero profits in each activity.

4) any output produced in excess of requirements and any resource not fully utilized receive a zero shadow price.

Valuation of resources and final products in terms of shadow prices that satisfy these conditions results in zero profits in all activities, an analogue of the perfectly competitive equilibrium.

Kantorovich's early formulation is incomplete. The author does not state explicitly the nature of the dual, his multipliers are mere computational devices that cannot be accorded the full status of "dual variables," nor does he supply "a rigorous, completely specified algorithm" for the solution. (5) During the next few years, however, he improved on it, filling important gaps, perfecting the numerical method, and extending the scope of linear programming from micro- to macroeconomic applications. (6) By 1943 he had come to realize that the construction of the national economic plan itself may be viewed as a grandiose problem in linear programming, in which solving multipliers turn into a system of optimal prices attaching to all resources and commodities in the economy, including capital and land, which had customarily been treated as free goods in Soviet economic accounting. Kantorovich saw in the existence of such a system of valuations that would result in zero profits in all activities both a necessary and the sufficient condition for the optimality of a national short-run economic plan. And thus the Walras-Pareto model was rediscovered by a Soviet mathematician who most likely did not even know their names.

The discovery of linear programming by L. V. Kantorovich was not fortuitous. It was a logical expression of his lifelong attitude toward pure and applied mathematics. Both the original idea and his subsequent determination in the face of highly unfavorable odds to pursue its implications were in keeping with his belief in the importance of reality testing of abstract mathematical propositions. Applied mathematics, far from being a stepchild of pure theory, if pursued with a similar rigor and precision, yields results which are important in their own right and, in turn, may feed back into the stream of pure analytical thought insights that might otherwise remain hidden in the complexity of "the real world."

His previous work on the development of "pure" functional analysis, a relatively novel field of mathematics, and on its applications to computational mathematics conformed to these principles. Starting in 1933-35, he created a new departure in functional analysis, a general theory of ordered vector spaces, a class of linear systems in which any finite set of elements has well-defined boundaries. They were named K-spaces to honor the discoverer. (7) Although he thus pioneered in a highly abstract branch of pure mathematics, he would not subscribe to G. H. Hardy's famous quip that the supreme test of a beautiful mathematical theorem is its utter uselessness. He expressed his position in a prize-winning essay:

> Traditionally, functional analysis has been considered a purely theoretical discipline incapable of any practical applications. [My] purpose is ... to repudiate this tradition ... and prove that concepts and methods of functional analysis can be as successfully employed in the construction and study of effective algorithms for solution of mathematical problems as they have been in their theoretical exploration (8)

In keeping with this attitude Kantorovich coauthored the world's first textbook in the theory of computation and approximation, an early (1936) precursor of the postcomputer "flood" of such studies (9) and a monument to his lasting concern with instilling rigor and precision into a field traditionally neglected by pure theorists.

II

By setting up a new paradigm of optimal planning, Kantorovich's discovery set in train a scientific revolution in Soviet

economics. But a discovery does not an innovation make until
it is widely diffused, accepted, and applied. The gestation period of the revolution initiated by Kantorovich has turned out to
be inordinately long.

The early reception accorded Kantorovich's discovery in
Leningrad was quite enthusiastic. His first presentation of
linear programming on May 13, 1939, at Leningrad University
received high praise on the part of the assembled mathematicians. Two weeks later the university convened a special conference of business executives at which the author presented a
wide range of economic applications of his method and outlined
efficiency gains that could be expected from them. "The in-
dustrialists in attendance unanimously expressed a great interest in his study and requested immediate publication." (10) The
book went to press in a record time (July 27, 1939), but in a
disappointingly small printing of one thousand copies. Its edi-
tor, A. R. Marchenko, predicted that Kantorovich's discovery
"will play an extremely useful part in the development of our
socialist industry." (11)

But the diffusion process stopped then and there for almost
twenty years. Faced with Soviet planners' and economists'
"mathematicophobia" (to use Kantorovich's own apt expression), his pathbreaking study received no attention whatever:
not a single review and only one footnote referring to it could
be found in economic literature until the late 1950s. (12) Un-
daunted by this lack of interest, Kantorovich achieved substantial progress during the early 1940s, working out extensions of
linear programming methods and of their economic applications.
He presented his findings in a number of papers and lectures
delivered at meetings of the Academy of Sciences, and he published some of them that were related to mathematical but not
economic aspects of linear programming. The manuscript of
his well-known book Best Utilization of Economic Resources [13],
containing his model of a short-run national economic plan, became upon its publication in 1959 the chief vehicle of the mathematical revolution in Soviet economics. It had been completed
in the main as early as 1943 but was denied publication. In the

meantime, linear programming had been independently discovered in 1947 and developed in the USA by G. B. Dantzig,
M. K. Wood, T. J. Koopmans, and others, none of whom was
aware of Kantorovich's pioneering work. (13) Practical applications of linear programming were equally conspicuous by
their absence. Even the plywood trust appears to have missed
its historical chance. It was not until 1950 that a Leningrad
railroad-car factory applied the new method in solving an industrial layout problem.

In the words of his close co-workers, "for long years Kantorovich remained completely isolated in his research on optimal planning." (14) Interestingly enough, during the same
years he was receiving increasing recognition as a leading
mathematician for his work on noneconomic topics. He continued making creative contributions to such diverse fields as
functional analysis and optimal layout of minefields and the
theory of approximation and automated programming; he pioneered in the development of Soviet computers and their applications in scientific research and took out a number of patents
for computing devices.

In 1949 he was awarded a State Prize for his essay on the applications of functional analysis. Nothing illustrates better the
uncompromising official attitude toward mathematical economics
in those days than the fact that the concluding part of this prize-
winning essay, which dealt with applications of functional analysis to economic problems, and which was announced for publication in a forthcoming issue of the journal in which the first part
had been published, has never appeared. (15) It was not until
1958 that Kantorovich was elected a corresponding member of
the Academy of Sciences, and in 1964 an Academician.

In the meantime, the potential usefulness of mathematical
methods of economic research and planning came to be gradually recognized, and in 1959 his seminal study Best Utilization
of Economic Resources [13] finally appeared, with a sixteen
years' delay, to be followed by a reprint of the 1939 classic
(see [1]), which had become virtually unobtainable. Kantorovich could no longer complain of a lack of critical attention.

The new paradigm could no longer be ignored and became a bitterly contested issue, with the more traditional-minded economists taking a skeptical (and usually not too well informed) view of the matter.

But Kantorovich was no longer alone. During the preceding years he had become the recognized leader of a growing group of young mathematicians interested in linear programming and related optimization methods. In the late 1950s he established in Leningrad the country's first workshop in mathematical methods for economists. In 1960 the workshop was moved, along with its director, to the Siberian branch of the Academy of Sciences in Novosibirsk and expanded into a major research center for the study of problems of optimal planning, with its own journal. The award of the Lenin Prize to Kantorovich in 1965, which he shared with the chief architect of the mathematical revolution in Soviet economics, Academician V. S. Nemchinov, and with V. V. Novozhilov (who was said by some to have been the first man to have read and understood Kantorovich's 1939 monograph), finally legitimized the cause for which they were fighting and immunized L. V. Kantorovich from further attacks. (16)

In 1971 Kantorovich accepted his present position at the Institute for the Management of the National Economy in Moscow, an elite institution that trains future leading executives. As at Novosibirsk, he continues to make important contributions to the theory and practice of optimal planning. Not unlike Monge, whose famous problem he once solved and who was known as "the prince of teachers," Kantorovich is an inspiring teacher who has played a leading role in the training of mathematical economists in the Soviet Union. This responsible and arduous task had to be started virtually from scratch after three decades of neglect during which mathematical methods had been banned from the economists' curricula and from the pages of economic journals.

Kantorovich's proposals for optimal planning were criticized, among other things, on the grounds that they infringed on the planners' freedom of choice and endangered the high rates of economic growth achieved with the traditional arithmetical methods of resource allocation in physical terms in the Soviet command economy.

The first objection was hardly justified. The new paradigm left unchanged the traditional objective function of maximizing total output with a given product mix, thus preserving central planners' freedom of choice with respect to ends. Their choice of means, however, was now explicitly recognized as a process of constrained maximization. Shadow prices, while expressing scarcity constraints, provided planners with a powerful tool for selection of the efficient plan variant. The adoption of the new paradigm would thus reduce the extent of arbitrariness allowed in the choice of means but would also significantly improve the static efficiency of such choice. This left open the possibility of a conflict between "Growth and Choice," to use Peter Wiles's apt formulation, or between the objectives of static versus dynamic efficiency. This criticism is of a considerable theoretical and practical import, and Kantorovich's subsequent work has aimed chiefly, although not exclusively, at expanding and deepening his original analytical framework so as to come to grips with problems of long-run planning in a dynamic economy.

III

Linear programming was discovered, as we recall, in attempting to solve a short-run planning problem on the level of a firm. In Best Utilization of Economic Resources [13], where its scope was extended to a short-run national economic plan, the author recognized the importance of a dynamic model with nonlinearities but did not discuss its characteristics at any great length. It was not until "A Dynamic Model of Optimum Planning" (1964) (Essay 4) and "Optimal Models of Long-term Planning" (1965) [27] that he developed a multiperiod model of the national economic plan designed for optimization over time, of which the short-run model was just a one-period cross section. (17)

Alfred Marshall once said that time is the source of greatest difficulties in economic analysis. Programming models are no exception to this rule.

The common theme running through most of Kantorovich's economic writings since 1959 that are presented in this selection of essays is the analysis of the time dimension of economic planning: long-run programming of economic activity, optimization over time, growth models, specific difficulties and complications arising in dynamic optimal planning. The main difficulties facing Kantorovich in his attempts to adapt his short-run model to the analytical and practical requirements of long-run planning could be described as follows:

1) Shadow prices referred to a point of time, but long-run planning deals with intertemporal choices in which present and future costs and benefits are not directly comparable in terms of static valuations.

2) Kantorovich's formulation of linear programming is concerned with an optimal intensity of utilization of a given resource vector when the technology matrix is given. In a long-run plan these constraints are released so as to allow for growth over time in the supply of resources and for technological progress. The need arises for a growth model explaining the interrelationships between these and similar variables and the rate of economic growth.

3) Nonlinearities have to be allowed for.

4) The probabilistic nature of various components of the plan has to be recognized.

5) Long-run planning is concerned with the strategic choices determining the course of economic development, in particular with the rate of investment. Although in the practice of Soviet planning this strategic choice itself is the prerogative of the system's directors, an analysis of its implications can hardly be left out of a model of long-run planning. It may be noted in passing that the discussion of such politically sensitive issues by mathematical economists in the 1920s was instrumental in bringing about their downfall.

How does Kantorovich attempt to deal with these issues ?

L. V. Kantorovich and Optimal Planning

In the dynamic model (Essay 4) all shadow prices are dated, as are all factors and products to which they pertain. A variable considered in n different periods is therefore treated as n different variables. Shadow prices tend to change (as a rule, to decline) over the planning period. The efficiency condition for an optimal multiperiod linear model is now the existence of a system of dated multipliers (shadow prices) such that when applied to different periods' activities, they would result in zero profits in all activities.

Comparability of benefits and costs over time is assured by discounting future valuations to their present values. The role of the discount factor is performed by an efficient rate of interest called "the normative efficiency of investment," e_i, derived as the marginal productivity of capital from a Cobb-Douglas aggregate production function. The optimal plan is selected as the variant with the highest present value. The discount factor becomes the keystone of the whole system. If one may be allowed an irreverent comparison, it runs through the multiperiod plan the way a skewer does through the slices of a shish kebab: it holds it together and gives the plan its Gestalt by assuring coherence and consistency of intertemporal choices.

In earlier writings a uniform rate of discount is assumed so as to provide a firm framework for the multiperiod system of shadow prices. In later years (e.g., in 1972 [65]) the author concedes that when probabilities of effectiveness of investment vary from industry to industry, the setting of a lower norm for, say, industry i would represent a justified allowance for the higher probability of payoff in that industry. Here, as on some other important occasions, Kantorovich moves away from a strictly deterministic to a probabilistic approach to economic planning.

The existence of nonlinearities in the real world is the source of considerable complications in planning models. Ever since 1940 Kantorovich has recognized their existence but has held that they could be treated with the analytical tools of a linear model. He then decided that the effects of nonlinearities occurring on the enterprise level would tend to be roughly offset when the number of enterprises is very large, as in the economy as a whole and over a sufficiently long period of time. In Essay 4 (1964) he proposed that nonlinear cost functions could be incorporated into the linear model by breaking cost curves down into linear segments and treating each segment as a separate ingredient of the plan. In the presence of nonlinearities shadow prices would be applicable only to differential changes in variables, and the existence of an efficient price system is no longer the sufficient condition for the existence of an optimal plan (although it still remains a necessary condition). Total returns to factors may, in the presence of nonlinearities, exceed total valuation of output in terms of shadow prices. This would either justify subsidy payments, which Kantorovich views as a compensation for external economies, or actual prices would have to be set at a higher level than optimal valuations.

Another difficulty is related to the level of aggregation of the model. Theoretically, nonlinearities, externalities, probabilistic factors, and similar complications can often be incorporated into the model by increasing the number of its variables and constraints. In his later writings Kantorovich becomes increasingly skeptical about the practical feasibility of such theoretically valid solutions and warns against the assumption that the number of variables in the model can be multiplied at will so as to take account of the multiplicity of characteristics of the real economy. "Such a hypothesis ... is very remote from reality" (18), due to the difficulty of solving problems with more than a few hundred constraints. In practice, he holds, the need for operational, stable models may favor a high degree of aggregation even if analytically the model would gain from disaggregation.

Kantorovich's analysis of economic growth in a centrally planned economy illustrates this attitude. Ever since his first growth model (Essay 5, 1959), he has operated with the assumption of an extremely aggregated, one-good, one-sector economy, using a Cobb-Douglas production function. Despite its lack of realism, the advantage of such a model in a centrally planned economy is, as pointed out by G. M. Heal, that it "makes it possible to concentrate on formulating the objective of [long-term] planning without considering complications which for this purpose are extraneous." (19) Gradually, in subsequent writings Kantorovich incorporates additional variables and studies the effects on the rate of economic growth of such factors as time lags, technological progress, depreciation methods, and others. Without mechanically transposing his conclusions to a multi-sector model, he manages to isolate what he considers to be the key variables in long-run planning. "The normative efficiency of investment," derived from his Cobb-Douglas production function, is the key link between static and dynamic efficiency, between planning for efficiency and planning for growth.

Kantorovich increasingly recognizes the uncertainty of the future in his discussion of long-run planning, although not in his mathematical models. (20) He provides a new solution to the old debate between the geneticists (Kondratiev, Bazarov) and teleologists (Strumilin, Stalin), who had argued whether an economic plan was a forecast or a command. According to Essay 14 (1974), [65] (1972), and other recent pronouncements, it contains both deterministic and probabilistic components, the latter especially in such areas as the planning of innovations and of mineral exploration and foreign trade. Kantorovich adds a novel twist to the argument by showing how the denial of uncertainty can actually lead to a conservative bias in planning (a charge once levied against geneticists!), e.g., by discriminating against investment in risky but promising innovations "if one demands that they result in a guaranteed, instantaneous payoff." (21)

A final point. The practice of Soviet economic planning, with its paradigm of maximizing output almost regardless of cost, resulted in building a highly effective machine for resource mobilization rather than for efficient resource utilization. Kantorovich's short-run model was concerned with static efficiency that traditionally tended to be treated as a secondary objective. His discussion of long-run planning, on the other hand, was bound to infringe on some sensitive issues of dynamic efficiency that touched upon the very core of the planners' preferences, such as the rationality of the strategy of extensive growth itself.

As essays included in this volume illustrate, Kantorovich does not emphasize these issues but neither does he stay away from them. He points to the limits of economic growth that can be achieved by merely increasing labor force and capital stock. He repeatedly emphasizes the importance of an incentive system well coordinated with the system of plan targets. He also explores a number of assumptions concerning the sensitive choice of the consumption-income ratio, C/Y:

1) it is given exogenously on the basis of noneconomic considerations, such as the external situation of the country, etc.;

2) several variants of the C/Y ratio are computed and then the optimal variant is selected that is also found to be politically acceptable;

3) a lower limit is set on the planned C/Y ratio by the effects of its decline on labor productivity, via real wages, and hence on the rate of growth of national income.

Kantorovich's treatment of the consumption issue, circumspect as it is, illustrates an important difference between priority settings in Western and in Soviet-type economies. His one-sector one-good model resembles Western "cake-eating" models of long-run planning, where a key concern is with providing for an optimum terminal capital stock. Yet he is not concerned with the possibility that the current consumption-income ratio may be in excess of the long-term optimum, but rather with the chance for an excessive sacrifice of present goods. "It often happens that in our concern for the future generations we harm the interests of the present one, even though these sacrifices may eventually turn out to have been unnecessary." (Essay 15)

In other words the real danger is not that of overeating but of excessive dieting — and then finding that the unconsumed cake, instead of serving the purpose of "starving to greatness," has to be thrown away.

IV

Essays presented in this volume are of a twofold interest: as contributions to the theory of economic planning and to our understanding of the Soviet economic system. Mathematical models are the cutting edge of modern economic analysis. In the hands of so skillful a practitioner as Academician Kantorovich, attempts to construct and use them yield revealing insights into important interrelationships, deep-seated maladjustments, and reservoirs of future strength of the Soviet economy.

The essays have been arranged in the following order. The first three items represent Kantorovich's assessment of the current status of optimal planning in the Soviet economy, of its progress, and of the key tasks awaiting solution.

The bulk of the essays, 4 through 14, deal with the determination of the rate of growth of national income and of the normative efficiency of investment in a centrally planned economy. The reader can follow the evolution of the author's views from the simple models of dynamic planning (Essay 4, 1964) and of economic growth (Essay 5, 1959) to his present position.

Using the conventional Western notation, let i equal normative efficiency of investment, Y equal national income, L equal employment, K equal capital stock, and C equal consumption. Then, from Essay 6 (1967) and assuming a Cobb-Douglas production function,

$$i = \frac{\dfrac{1}{Y} - \dfrac{dY}{dt}\dfrac{L'}{L}}{1 - \dfrac{C}{Y} - \dfrac{L'K}{LY}}$$

In Essay 8 (1970) further variables are added, expressing the effects of technological progress and of depreciation policies. The latter reflect both physical wear and tear and an allowance for obsolescence, a controversial issue in Soviet practice. In Essay 11 (1973) further refinements are added. The rate of growth of income is now to be maximized at each instant of time, and new lagged variables are introduced to account for the effect on the maximand of output capacities of new investment projects being put into operation, of their capital intensity, and of the time required to complete their construction and to achieve a full-capacity operation. The interest of the theoretical discussion is enhanced by the fact that Kantorovich acts as his own econometrician. Of special interest is Essay 8, where he feeds into his model statistical data and estimates in order to determine the actual numerical values of its parameters for the Soviet economy in 1965. The efficient rate of interest was then estimated at roughly 22 percent. When the effects of technological progress and other additional variables, mentioned above, are taken into account in Essay 11, this estimate is reduced to roughly 18 percent. It may be recalled that the author goes beyond the normal assumptions of Pareto optimality and allows for intersectoral variation of this norm where warranted by the risk factor, etc.

In addition to grandiose problems of what William Baumol calls "the magnificent dynamics," Kantorovich has never hesitated to tackle specific issues of a more narrow scope that arise in the process of devising and implementing optimal planning in various areas of the Soviet economy.

Essay 13 (1966) is representative of the author's efforts to formulate principles of an optimal replacement theory and thus bridge the existing gap between new and old capital assets in Soviet investment policies. (22) Depreciation allowances are shadow prices that should fully reflect the real factor cost of capital assets, including obsolescence, and assist in an optimal sequencing of replacement decisions. One portion of the annual write-off should vary with the asset's lifetime, another with its use time. Higher write-offs during the asset's early life help "insure" against the probability of obsolescence through technological progress, while lower (or zero) write-offs on old machines discourage their premature scrapping.

The main point of Essay 15 (1971), patiently reiterated by the Soviet economist on numerous occasions, is that although the pressing need is for further progress in dynamic planning, even the lessons of his static model have not yet been learned in the practice of Soviet price formation. He shows the cost to the economy of irrational pricing and benefits that can be expected from bringing actual factor prices (especially land and capital charges) in line with their relative real scarcities. (23)

In Essay 16 (1972) Kantorovich discusses current requirements for computer hardware needed for implementation of optimal planning and day-to-day management of the national economy, and in Essay 17 (1972) he takes stock of the needs for computer software. Criteria for the selection of models, programs, and algorithms are discussed.

In Essay 18 (1970) the author reports on a unique project successfully carried out under his direction during the 1960s and originally proposed by him as early as 1940: to find an optimal production scheduling and product distribution for rolled metal and tubing for the entire Soviet steel industry, with exogenously given demand and supply estimates and with a ceiling on the transport input. A general solution would require a matrix on the order of three million by thirty thousand, Kantorovich remarks, and "nobody had ever solved a problem of this magnitude." (Essay 18) It took the combined efforts of a number of research teams and over eight years to arrive at a solution.

V

This brings us a long way from the original plywood trust problem three decades earlier. Throughout this long progress, which was to be crowned with a Nobel Prize in Economics in 1975, Kantorovich has displayed an admirable unity of purpose, consistency of method in achieving his breakthroughs, and persistence in promoting their implementation. In his work on optimal planning, as in mathematics, he combines the rigorous pursuit of "long chains of deductive reasoning" with a keen sense for the constraints imposed by the complexity of economic reality. Especially in his recent economic writings, he tends to synthesize rather than dichotomize.

Thus although a passionate proponent of factor-cost pricing, he now allows for the use of subsidies and of a nonuniform rate of interest. He warns against identifying optimal planning with either market socialism (see Essay 1, p.3) or with the Laplacian pipe dream of a perfectly centralized deterministic planning. In an ingenious simile he draws an analogy between an economic system and a space rocket. In simple cases its course can be preprogrammed, an analogue of deterministic planning in a simple economic system; but more complex tasks require continuous revisions of the program through feedback loops. Analogously, planning a complex, modern economy presupposes "a combination of a [central] plan with autonomous actions by decentralized units." (24) He warns (unlike market socialists of Oskar Lange's early persuasion) that it is not enough to merely inform such units what course of action is most beneficial to the economy: incentives to follow it must also be provided.

In a similar vein he proposes to use a combination of directive and indicative planning, although "there is no answer yet as to...how one can assure an optimal combination of direction with forecasting." (25) Central allocation of supplies is to be used in conjunction with a free wholesale trade in producer goods. (26)

This pluralistic approach is in keeping with Kantorovich's cautious conclusions concerning the role of mathematical models in economic analysis. He points to "a limited, conditional, and approximate character of inferences that can be drawn from models of linear programming..." and warns against "a dogmatic, noncritical acceptance of findings derived from the use of models, against their absolutization and the overrating of the method in question." (27) In brief, he sees a danger of their being elevated to a rank of a new dogma, unless they are used in a critical, creative, and nonformalistic manner.

The significance of Kantorovich's theory of optimal planning can not yet be judged by its practical applications. Important as these may be, they lag significantly behind theory. Its main significance may lie, at least for the time being, elsewhere.

In the West the discovery of linear programming by Dantzig gave rise to significant progress in economic analysis but not to a scientific revolution. In the Soviet Union Kantorovich's discovery, while important in its own right, served the additional role of a vehicle for introduction and explicit recognition of a logic of choice under constraints. The new paradigm of optimal planning has served an important role in laying bare the foundations of economic analysis for the new generations of Soviet economists. The stage of enlightenment in this seminal process is far from complete, while the extremely promising stage of practical results is barely setting in. Kuhn's reminder is apposite: "Paradigm debates are not really about relative problem-solving ability.... A decision between alternate ways of practicing science is called for, and in the circumstances that decision must be based less on past achievement than on future promise." (28)

Notes

1) Biographical information in this essay was derived mainly from the following sources: Uspekhi matematicheskikh nauk, 1962, XVII, no. 4, pp. 200-15; op. cit., 1972, XXVII, no. 3, pp. 221-27; Optimizatsiia, 1971, no. 3 (20), pp. 7-13; Bol'shaia sovetskaia entsiklopediia, 3rd ed., vol. 11, p. 1007.

2) See [1], pp. 5-6.

3) Gaspard Monge (1746-1818) was the inventor of descriptive geometry.

4) T. J. Koopmans, "A Note about Kantorovich's Paper 'Mathematical Methods of Organizing and Planning Production,'" Management Science, 1960, 6, no. 4, pp. 364-65.

5) A. Charnes and W. W. Cooper, "On Some Works of Kantorovich, Koopmans, and Others," Management Science, 1962, 8, no. 3, pp. 246, 249, 251.

6) Much of this work was presented at conferences but remained unpublished or published with considerable delays. Thus the algorithm was supplied in [5], a paper completed as early as 1940 but not published until 1949; the relationship between the primal and the dual was brought out in [3] (1942) and in [8], which was presented as a paper in May of 1941 but not accepted for publication until 1957. Some of the economic applications and numerical solutions, such as [8] and [9], appeared with an eight to nine years' delay.

7) "L. V. Kantorovich," Uspekhi matematicheskikh nauk, 1962, XVII, no. 4, p. 206.

8) L. V. Kantorovich, "Funktsional'nyi analiz i prikladnaia matematika," Uspekhi matematicheskikh nauk, 1948, III, no. 6, p. 89.

9) L. V. Kantorovich and V. I. Krylov, Priblizhennye metody vysshego analiza, 1st ed., 1936; 5th ed., 1962, p. 9.

10) A. R. Marchenko, preface to [1], p. 1.

11) Ibid.

12) The footnoter was Professor V. V. Novozhilov, who was to share with Kantorovich and Nemchinov a Lenin Prize in 1965. See his "Metody nakhozhdeniia minimuma zatrat v sotsialisticheskom khoziaistve," Trudy Leningradskogo politekhnicheskogo instituta, 1946, no. 1, p. 336.

13) It was not until 1958 that Koopmans received a copy of [1] from the author, upon learning by chance of its existence. See Koopmans, "Note," p. 364.

14) V. L. Makarov and G. Sh. Rubinshtein, "O vklade L. V. Kantorovicha v razvitie ekonomicheskoi nauki," Optimizatsiia, 1971, no. 3 (20), p. 11.

15) Kantorovich, "Funktsionalnyi analiz," p. 90.

16) The immunity seems to wear out. Just before the award of the Nobel Prize to Kantorovich, B. Griaznov published a criticism of linear programming quite in the spirit of the early, uninformed attacks. Its usefulness in optimal macroeconomic planning is questioned and, in particular, the interpretation of "solving multipliers" as scarcity prices is denied validity. See B. Griaznov in Planovoe khoziaistvo, 1975, no. 9, pp. 155 ff.

17) See [24], p. 334.

18) See [64], p. 172.

19) G. M. Heal, The Theory of Economic Planning, North Holland, 1973, p. 257.

20) Although he has the required expertise as author of a text on the theory of probability: see his Teoriia veroiatnostei, 1946.

21) See [65].

22) See especially [28] (1965), [36] (1966), [57] (1971).

23) See [67], p. 14.

24) See [64], p. 206.

25) Ibid., p. 179.

26) Ibid., p. 208.

27) Ibid., pp. 181-82.

28) Thomas S. Kuhn, The Structure of Scientific Revolutions, 2nd ed., The University of Chicago Press, 1970, pp. 157-58.

Selected Economic Writings of L. V. Kantorovich

[1]   Matematicheskie metody organizatsii i planirovaniia proizvodstva. Leningrad: Leningrad University Press, 1939, 68 pp. (reprinted with minor changes in Nemchinov, V. S., ed., Primenenie matematiki v ekonomicheskikh issledovaniiakh, Vol. I, Moscow, 1959, pp. 251-309).

[2]   "Ob odnom effektivnom metode resheniia nekotorykh klassov ekstremal'nykh problem." DAN SSSR, Vol. 28, No. 3 (1940), pp. 212-15.

[3]   "O peremeshchenii mass." DAN SSSR, Vol. 37, No. 7-8 (1942), pp. 227-29.

[4]   "Ob odnoi probleme Monzha." Uspekhi matematicheskikh nauk, Vol. 3, No. 2 (1948), pp. 225-26.

Selected Writings of L. V. Kantorovich

[5] (with M. K. Gavurin) "Primenenie matematicheskikh metodov v voprosakh analiza gruzopotokov," Problemy povysheniia effektivnosti raboty
transporta. Moscow, 1949, pp. 110-38 [written in 1940].

[6] "Podbor postavov obespechivaiushchikh maksimal'nyi vykhod produktsii
pri zadannom assortimente." Lesnaia promyshlennost', No. 7 (1949),
pp. 15-17, and No. 8 (1949), pp. 17-19 [written in 1941].

[7] (with V. A. Zalgaller) Raschet ratsional'nogo raskroia promyshlennykh
materialov. Leningrad, 1951, 198 pp.; 2nd ed., Moscow, 1971, 300 pp.

[8] "O metodakh analiza nekotorykh ekstremal'nykh planovo-proizvodstven-
nykh zadach." DAN SSSR, Vol. 115, No. 3 (1957), pp. 441-44 [findings
first reported in a paper given at Leningrad University, May 12, 1941].

[9] (with G. Sh. Rubinshtein) "Ob odnom funktsional'nom prostranstve i nekotorykh ekstremal'nykh zadachakh." DAN SSSR, Vol. 115, No. 6 (1957),
pp. 1058-61.

[10] "Vozmozhnosti primeneniia matematicheskikh metodov v voprosakh pro-
izvodstvennogo planirovaniia," Organizatsiia i planirovanie ravnomernoi
raboty mashinostroitel'nykh predpriiatii. Moscow, 1958, pp. 338-53.

[11] (with G. Sh. Rubinshtein) "Lineinoe programirovanie," Bol'shaia sovetskaia entsiklopediia. Vol. 51, 2nd ed., 1958.

[12] (with A. A. Ivanov) "Operatsii issledovaniia," Bol'shaia sovetskaia entsiklopediia.

[13] Ekonomicheskii raschet nailuchshego ispol'zovaniia resursov. Moscow:
AN SSSR, 1959, 344 pp.; 2nd ed., 1960.

[14] "Dal'neishee razvitie matematicheskikh metodov i perspektivy ikh primeneniia v planirovanii i ekonomike," V. S. Nemchinov, ed., Primenenie
matematiki v ekonomicheskikh issledovaniiakh. Vol. I, Moscow, 1959,
pp. 310-53.

*[15] (with L. I. Gor'kov) "O nekotorykh funktsional'nykh uravneniiakh vozni-
kaiushchikh pri analize odnoproduktovoi ekonomicheskoi modeli." DAN
SSSR, Vol. 29, No. 4 (1959), pp. 732-35.

[16] "O primenenii sovremennykh matematicheskikh metodov pri opredelenii
ekonomicheskoi effektivnosti kapital'nykh vlozhenii," Ekonomicheskaia
effektivnost' kapital'nykh vlozhenii i novoi tekhniki. Moscow, 1959,
pp. 227-37.

[17] "Vystuplenie na sessii Akademii Nauk SSSR." Vestnik Akademii nauk
SSSR, No. 4 (1959), pp. 59-61.

[18] "Ob ischislenii proizvodstvennykh zatrat." Voprosy ekonomiki, No. 1
(1960), pp. 122-34.

[19] "Optimal'noe planirovanie i ekonomicheskie pokazateli," Trudy nauchnogo soveshchaniia o primenenii matematicheskikh metodov v ekonomicheskikh issledovaniiakh i planirovanii. Vol. I, Moscow, 1961, pp. 67-99.

[20] "Vystupleniia," Trudy nauchnogo soveshchaniia o primenenii matematicheskikh metodov v ekonomicheskikh issledovaniiakh i planirovanii. Vol. I,
Moscow, 1961, pp. 160-64, 259-71.

[21] "O nekotorykh novykh podkhodakh k vychislitel'nym metodam i obrabotke

*Essay appears in translation in this volume.

nabliudenii." Sibirskii matematicheskii zhurnal, Vol. 3, No. 3 (1962), pp. 701-9.

[22] "Pobedy elektronnykh ekonomistov." Komsomol'skaia pravda, July 7, 1962, p. 4.

[23] "Usloviia optimal'nogo planirovaniia." Ekonomicheskaia gazeta, April 20, 1963, pp. 7-8.

*[24] "Dinamicheskaia model' optimal'nogo planirovaniia," N. P. Fedorenko, ed., Planirovanie i ekonomiko-matematicheskie metody. Moscow, 1964, pp. 323-45.

[25] "Perspektivy primeneniia matematicheskogo optimal'nogo planirovaniia v sel'sko-khoziaistvennom proizvodstve," Primeneniia matematiki v ekonomike sel'skogo khoziaistva. Moscow, 1964.

[26] Matematicheskie problemy rascheta i analiza optimal'nykh dinamicheskikh modelei, Novosibirsk, 1965, 11 pp.

[27] (with V. L. Makarov) "Optimal'nye modeli perspektivnogo planirovaniia," V. S. Nemchinov, ed., Primenenie matematiki v ekonomicheskikh issledovaniiakh. Vol. III, Moscow, 1965, pp. 7-87.

[28] (with I. V. Romanovskii) "Amortizatsionnye platezhi pri optimal'nom ispol'zovanii oborudovaniia." DAN SSSR, Vol. 162, No. 5 (1965), pp. 1115-18.

[29] "Na osnove matematicheskikh metodov." Ekonomika stroitel'stva, No. 3 (1965).

[30] "Matematika i ekonomika." Pravda, August 24, 1965.

[31] "Printsip optimal'nosti." Ekonomicheskaia gazeta, No. 45 (1965).

*[32] "Amortizatsionnye otchisleniia i otsenka effektivnosti novoi tekhniki v sisteme optimal'nogo planirovaniia," Matematiko-ekonomicheskie problemy, Vol. 58, Trudy Leningradskogo inzhenerno-ekonomicheskogo instituta, 1966, pp. 3-11.

[33] "Matematicheskie metody v reshenii khoziaistvennykh zadach." Kommunist, No. 10 (1966), pp. 64-74.

[34] "Razvitie matematicheskikh metodov ekonomicheskogo analiza." Vestnik Akademii nauk SSSR, No. 10 (1966), pp. 9-15.

[35] (with A. G. Pinsker) "Matematicheskaia podgotovka ekonomistov." Vysshaia shkola, 1966.

[36] "Struktura amortizatsionnykh ischislenii pri statsionarnoi nagruzke mashinnogo parka." DAN SSSR, Vol. 166, No. 2 (1966), pp. 309-12.

[37] "Matematicheskie metody optimal'nogo planirovaniia," Matematicheskie modeli i metody optimal'nogo planirovaniia, Novosibirsk, 1966, pp. 109-16.

*[38] "Matematicheskie optimal'nye modeli v planirovanii razvitiia otrasli i tekhnicheskoi politike." Voprosy ekonomiki, No. 10 (1967), pp. 102-15.

*[39] (with A. L. Vainshtein) "Ob ischislenii normy effektivnosti na osnove odnoproduktovoi modeli razvitiia narodnogo khoziaistva." Ekonomika i matematicheskie metody, Vol. 3, No. 5 (1967), pp. 697-710.

[40] "Ob ischislenii normy effektivnosti na baze odnoproduktovoi modeli

*Essay appears in translation in this volume.

razvitiia khoziaistva." Optimal'noe planirovanie, No. 8 (1967), pp. 37-51.

*[41] (with I. G. Globenko) "Odnoproduktovaia dinamicheskaia model' pri nalichii mgnovennoi prevrashchaemosti fondov." DAN SSSR, Vol. 174, No. 3 (1967), pp. 18-21.

*[42] (with I. G. Globenko) "Dinamicheskaia model' ekonomiki." DAN SSSR, Vol. 176 (1967), pp. 997-98.

[43] (with A. B. Gorstko) "Matematika i ekonomika," Nauka i chelovechestvo, 1967.

[44] (with V. L. Makarov) "Voprosy razrabotki i ispol'zovaniia krupnoaggregirovannoi modeli optimal'nogo planirovaniia." Optimal'noe planirovanie, No. 8 (1967), pp. 23-35.

[45] (with A. B. Gorstko) Matematicheskoe optimal'noe programmirovanie v ekonomike. Moscow, 1968.

[46] "O putiakh dal'neishego sovershenstvovaniia tsenoobrazovaniia," V. P. Diachenko, ed., Itogi reformy tsen i perspektivy tsenoobrazovaniia. Moscow, 1968, pp. 35-40.

[47] "Ob ispol'zovanii matematicheskikh modelei v tsenoobrazovaniiu na novuiu tekhniku." Sovershenstvovanie tsenoobrazovaniia i nauchno-tekhnicheskii progress. Moscow, 1968, pp. 46-50.

[48] "Vystuplenie." Materialno-tekhnicheskoe snabzhenie, No. 10 (1969), pp. 68-71.

[49] (with Ia. I. Fet) "O vozmozhnosti povysheniia proizvoditel'nosti universal'noi tsentral'noi vychyslitel'noi mashiny pri reshenii ekonomiko-matematicheskikh zadach." Ekonomika i matematicheskie metody, Vol. 5, No. 2 (1969), pp. 276-79.

[50] (with V. N. Bogachev) "Tsena vremeni." Kommunist, No. 10 (1969).

[51] "Puti primeneniia matematicheskikh metodov v sel'skom khoziaistve." Optimal'nye modeli orosheniia. Moscow, 1969.

*[52] "Opyt optimal'noi zagruzki prokatnykh stanov." Materialno-tekhnicheskoe snabzhenie, No. 4 (1970), pp. 87-91.

[53] "Metody optimizatsii i matematicheskie modeli ekonomiki." Uspekhi matematicheskikh nauk, Vol. 25, No. 5 (1970), pp. 107-9.

*[54] (with A. L. Vainshtein) "Esche ob ischislenii normy effektivnosti na osnove odnoproduktovoi modeli razvitiia narodnogo khoziaistva." Ekonomika i matematicheskie metody, Vol. 6, No. 3 (1970), pp. 407-15.

*[55] (with V. N. Bogachev and V. L. Makarov) "Ob otsenke effektivnosti kapital'nykh zatrat." Ekonomika i matematicheskie metody, Vol. 6, No. 6 (1970), pp. 811-26.

[56] (with V. L. Makarov) "Differentsial'nye i funktsional'nye uravneniia voznikaiushchie v modeliakh ekonomicheskoi dinamiki." Sibirskii matematicheskii zhurnal, Vol. 11, No. 5 (1970), pp. 1046-59.

[57] "Struktura amortizatsionnykh platezhei v nekotorykh modeliakh ispol'-zovaniia mashinnogo parka." Optimizatsiia (Novosibirsk), Vol. 21, No. 4 (1971), pp. 7-20.

[58] (with M. I. Virchenko) "Matematiko-ekonomicheskii analiz planovykh

*Essay appears in translation in this volume.

reshenii i ekonomicheskie usloviia ikh realizatsii," L. V. Kantorovich and V. P. Mozhin, eds., Voprosy analiza planovykh reshenii v sel'skom khoziaistve, Vol. I, Novosibirsk, 1971, pp. 5-40.

*[59] "O tsenakh, tarifakh i effektivnosti ekonomiki." Ekonomika i organizatsiia promyshlennogo proizvodstva, No. 1 (1971), pp. 25-30.

*[60] "Puti razvitiia vychislitel'nykh sredstv dlia resheniia bol'shykh zadach optimal'nogo planirovaniia i upravleniia." Optimizatsiia (Novosibirsk), No. 6 (1972), pp. 5-7.

[61] (with A. V. Gorstko) Optimal'nye resheniia v ekonomike. Moscow, 1972, 232 pp.

*[62] (with E. G. Gol'shtein, V. L. Makarov, and I. V. Romanovskii) "Matematicheskie problemy optimal'nogo planirovaniia i upravleniia." Vestnik Akademii nauk SSSR, No. 10 (1972), pp. 70-79.

[63] "O matematicheskom obespechenii ASU Metall." Pribory i sistemy upravleniia, No. 12 (1972), pp. 8-10.

*[64] (with V. I. Zhiianov) "Odnoproduktovaia dinamicheskaia model' ekonomiki, uchityvaiushchaia izmenenie struktury fondov pri nalichii tekhnicheskogo progressa." DAN SSSR, Vol. 211, No. 6 (1973).

*[65] "Ekonomicheskie problemy nauchno-tekhnicheskogo progressa." Ekonomika i matematicheskie metody, Vol. 10, No. 3 (1974), pp. 432-48.

*[66] "Optimal'noe planirovanie: nereshennye zadachi." Ekonomika i organizatsiia promyshlennogo proizvodstva, No. 5 (1974), pp. 3-9.

*[67] "Dostizheniia ekonomicheskoi nauki — v praktiku." Ekonomicheskaia gazeta, No. 26 (1974), p. 14.

*[68] "Matematicheskie metody — ekonomike." Literaturnaia gazeta, October 22, 1975 (No. 43), p. 13.

---

*Essay appears in translation in this volume.

L.V. Kantorovich:
Essays in
Optimal Planning

1

Mathematical Methods in Economics*

L. V. KANTOROVICH

Academician L. V. Kantorovich, distinguished Soviet scientist and head of the task force for economic-mathematical methods and operations research at the Institute for the Management of the National Economy, and Professor T. Koopmans, an American specialist in mathematical economics, were awarded the Nobel Prize in Economics in 1975 for their contribution to the theory of optimal resource utilization. Literary Gazette [Literaturnaia gazeta] correspondent O. Moroz met with Academician Kantorovich and asked him to answer several questions.

Q. Leonid Vital'evich, ten years ago you and your colleagues — Academician V. S. Nemchinov and Professor V. V. Novozhilov — were awarded the Lenin Prize, the highest award that our country confers on a scientist's work, for the scientific elaboration of the method of linear programming and economic models. This work has now attracted the attention of the Swedish Academy of Sciences. Would you tell the readers of Literary Gazette about the essence of your many years of research?

A. I began research on the theory of optimal resource utilization, for which I was awarded the Nobel Prize together with

---

*Literaturnaia gazeta, no. 43, October 22, 1975.

the American scientist Professor Koopmans, at the end of the 1930s, when I was working at Leningrad University and in the Leningrad division of the Mathematical Institute. Then, for the first time, I turned my attention to such practical problems as the optimal distribution of work between machine tools, the most rational layout pattern for metal, the optimal use of means of transport and of sown areas, etc. All of these topics belong to the same group of mathematical problems — so-called extremal problems. It became apparent that it was practically impossible to solve these problems by classical mathematical methods since it would have been necessary to solve tens of thousands or even millions of systems of equations. For this reason new methods were developed that were later called the methods of linear programming, and after further development — the methods of mathematical optimal programming. The principles of these methods were set forth in my brochure Mathematical Methods in the Organization and Planning of Production [Matematicheskie metody organizatsii i planirovaniia proizvodstva], which was published by Leningrad University in 1939.

It subsequently became apparent that the methods of linear programming could be extended to the solution of problems of a broader scale — problems pertaining to the planning of freight flows, the rational utilization of capacities on the scale of the branch, and subsequently, problems of national economic planning as a whole. Of course, under the conditions of a socialist economy.

Mathematical models of these problems were built in the process of such solution. As a result, it not only became possible to calculate the plan, but there also emerged a base for the scientifically substantiated calculation of a number of key economic indicators: the price of rent and indicators of the economic effectiveness of capital investments.

These works received their first culmination in the book Economic Calculation of Optimal Utilization of Resources [Ekonomicheskii raschet nailuchshego ispol'zovaniia resursov], which was published by the USSR Academy of Sciences in 1959. They subsequently underwent significant development in the works of many younger Soviet economists, particularly at the Central Economic-Mathematical Institute, at the Institute of Economics, at the Institute of Mathematics of the Siberian Branch of the USSR Academy of Sciences, at the Institute of Cybernetics of the Ukranian Academy of Sciences, and many other scientific institutes and higher educational institutions.

In recent years the same methods have been successfully used to resolve problems pertaining to such comparatively new areas as the economics of natural resource utilization and environmental protection, many problems relating to the economics of technical progress, etc.

At the same time, it must be said that a large number of important problems in this area remain unsolved. Nor is this surprising if one considers the extraordinary complexity and diversity of "economic matter." We recall that both in physics and in mechanics, where mathematics has been applied for hundreds of years, there are many unsolved problems associated with the application of mathematics.

Q. What practical application have your ideas found in the national economy?

A. The methods of optimal mathematical programming were first applied in practice about twenty-five years ago. Thus at the Leningrad Egorov Railroad Car Plant they were used for the rational laying out of material and produced a considerable economic effect. These methods have been in use for more than fifteen years for solving transport problems and planning freight flows and truck routing in Moscow, Leningrad, and other cities. This has made possible a considerable reduction in the volume of the work and the more useful utilization of transport.

Many other examples of the application of mathematical methods of optimization could be cited. Calculations of optimal long-range plans have been made for many branches, and this has led to a considerable savings in capital investments. These calculations have found a definite use in planning.

Q. What, in your view, are the prospects for the application of mathematical methods of optimization in the national economy?

A. In the future these methods may find even broader application. The Party and the government are continuously pointing us in this direction. Thus the Twenty-Fourth Congress of the CPSU posed the target of "securing the broad application of economic-mathematical methods in the interests of improving the planning of the national economy and management...."

The application of these methods has received particularly great recognition in long-term planning. It strikes me as very important to make use of these methods in current planning as well. It is specifically in the latter instance that (1) they may produce an immediate and indisputable effect, and (2) since current planning makes much more rigid demands on decisions, it will make it possible to improve, to "polish" the methods of optimal planning and also to raise their authority. Optimization methods must be used more widely in the study of consumption and in questions pertaining to the further raising of the standard of living.

Mathematical methods and models of current planning should occupy a larger place in the elaboration of automatic control systems, which, as we know, are presently receiving a great deal of attention. The greatest effect from these systems can be obtained not so much as a result of the solution of purely informational problems, not so much from the automation of record keeping, as from the introduction of elements of optimization into current and long-term planning. This will not only increase the effectiveness of managerial work but will also raise the level of return in production proper, the level of return on capital, labor productivity in basic production, etc.

Naturally, however, for such broad use of mathematical methods of optimization, the efforts of only scientific workers and scientific institutes alone are not enough. It is essential that industrial personnel take a direct part in this work. This work should be headed by the country's major economic planning agencies, such as Gosplan, Gossnab, the Committee on Prices, the Committee on Labor and Wages, the Ministry of Finance, as well as the Committee on Science and Technology. Without their participation and monitoring, it appears that even projects that have been tested at a number of enterprises are not broadly disseminated for many years.

6

Q. How do you explain the fact that a work which derived
from the study of the economy of socialist society was awarded
a prize (for the first time, by the way) by a scientific organization of a capitalist country?

A. I see nothing surprising here since a considerable part
of the work represents the development of the general scientific
apparatus for investigating economic systems and economic objects which are, of course, found in the economy of any economically developed country. After all, as has already been
stated, these methods are applicable at the level of an individual
enterprise, at the level of an individual sector, etc. Thus it is,
to a certain degree, common scientific attainment. Confirmation of the universality of these methods is provided by the fact
that they were discovered independently, if somewhat later, in
the Western countries, and in particular in the works of Professor Koopmans (I take the opportunity to note also the contribution by Professor G. B. Dantzig, the American scientist,
and the Soviet researchers Professors V. V. Novozhilov and
A. L. Lur'e).

At the same time, it must be said — I noted this point in my
first work in 1939 — that optimal mathematical methods should
be considered most valuable and most suitable under conditions
of the socialist economic system, in which scientific planning
plays an immeasurably greater part. We can cite as an example the problem of placing orders for metal products on the
scale of the entire country, which is presently being resolved
by Gossnab using the methods of linear programming with our
direct participation. Of course, in any other — nonsocialist —
country it is inconceivable that the problem of utilizing hundreds of rolling and tube mills to fill the orders of tens of
thousands of different organizations could be solved simultaneously and harmoniously. This applies all the more to the national economy as a whole, the conscious scientific planning of
which is possible only under conditions of socialist society.

Here we must also note the following point. The use of mathematical models of the economy, while providing effective
means for improving the planning methods of economic manage-

ment, makes possible the rational combination of national planning decisions with local economic initiative while leaving the decisive role to the former. For this reason pronouncements appearing in the Western press (based on ignorance or deliberate distortion of the essence of the matter) that associate optimal planning with theories of "market socialism" are entirely without foundation. Criticizing similar theories in my book back in 1959, I wrote that their source is the underevaluation of successes already attained in the planning and economic development of socialist countries as well as the underevaluation of the great potential for the further improvement of economic planning inherent in the nature of the socialist mode of production — the most perfect in the history of mankind.

2

Let Us Apply the Achievements
of Economic Science*

L. V. KANTOROVICH

The enormous scale and profound interrelations of modern socialist production make the management of production more and more complex and important and require the continuous raising of its scientific level. This circumstance dictates the need for continuous progress and for the rapid development of economic science. The Communist Party is persistently enlisting the aid of economics scholars in resolving urgent problems in economic practice, provides daily aid to economic science, and concerns itself with its comprehensive development.

Soviet economists have made a definite contribution to improving the mechanism of management. Here, in particular, we should note the elaboration of methods of mathematical model-building of economic processes and the use of computers in the management of production. The elaboration of the methodology of optimal planning by Soviet scientists is of particular importance.

In the last ten years these and other new methods have been carefully tested in practice and have found quite broad application in long-term branch planning.

Of course, the advances of economic science still hardly

---

*Ekonomicheskaia gazeta, 1974, no. 26.

correspond to the demands made on it. At the same time, it would be incorrect to say nothing about shortcomings in the practical application of a number of economic elaborations that have already been tested and that have proven their validity and effectiveness.

Unutilized Opportunities

Thus in our opinion new mathematical methods of optimal planning are clearly not being sufficiently used in the economy. While long-term optimal planning calculations have been made for dozens of branches and have yielded a major calculated saving in a number of instances, they are virtually not used in current planning. And yet the application of optimization methods, e.g., for distributing the current production program between enterprises within a branch or subbranch can yield a great and immediate effect. This is clearly confirmed by a project done in the system of Gossnab and the USSR Ministry of Ferrous Metallurgy on the organization of the rational utilization of pipe and rolling mills. Unfortunately, there are literally only a handful of such examples. In the majority of cases automated branch management systems do not even incorporate optimal current planning elements, or else they have not been activated.

Evidently, the fact takes its toll that the application of mathematical methods in current planning is associated with a high degree of responsibility, with rigid deadlines for obtaining solutions, with the difficulty of obtaining information in good time, and with the necessity of correcting plans and interdepartmental coordination. Here it will be necessary to display persistence and initiative in order to overcome the obvious force of inertia and fear of the new.

Moreover, theoretical recommendations for improving a number of economic indicators hold great importance for practice. In many cases, however, they also still do not find complete utilization in planning organs, and without this their great effect for the national economy is not realized. Thus important principles of payment for capital and of taking into acount the capital intensiveness in prices that have been formulated by economic science as a result of a number of discussions are still not reflected with a sufficient degree of consistency in the system of price formation. In the process of establishing wholesale prices, the capital intensiveness of a specific product is not as a rule taken into account, which often leads to incorrect price relationships that put new, progressive types of items in especially disadvantageous positions.

Practically speaking, up until now rent, which is of great importance for the mining industry and for agriculture, has not been properly considered in price. The methodology of optimal planning provides sufficiently objective and effective methods for establishing prices that concretize the theory of rent with respect to modern conditions. The introduction of corresponding changes in purchase prices and in the system of economic indicators of agricultural production could play a definite part in the further economic strengthening of collective and state farms and in the creation of more favorable conditions for the intensification of agriculture, the specialization of farms, and the improvement of interfarm relations. It would serve to equalize economic conditions for agricultural enterprises situated in different natural zones and would promote the consistent observance of the principle of equal pay for equal work. The time has clearly come to conduct such an experiment, if only in a few regions or republics.

Science has substantiated in detail the necessity of taking into account the time factor in economic calculations. In principle this factor was recognized and reflected in the accepted methods for calculating the effectiveness of capital investments, which envisage the discounting of expenditures incurred at different times. But nonetheless, this principle is not implemented systematically and consistently in our economic practice. Sometimes, progressive projects that pay off in a much shorter time than envisaged in the norms do not find an application for years. The systematic accounting of intertemporal differences in con-

struction estimates and in the evaluation of the work of construction organizations would permit a sharp reduction in the
scattering of capital investments, would reduce construction
delays and gestation periods, and would greatly curtail the
freezing of resources in incomplete construction.

While the list of such examples could be continued, even
without doing so it is clear that the "freezing" or the incomplete, inconsistent utilization of the conclusions and recommendations of economic science occasion considerable national
economic losses and mean missed opportunities in increasing
the effectiveness of social production. And it is no accident
that the documents of the Twenty-Fourth Party Congress emphasize the obligation of economic managers to learn how to
manage in the new way, particularly on the basis of the application of the methods of economic-mathematical model-building,
systems analysis, and modern computer technology.

Unfortunately, this important instruction has hardly been
fully realized. It is often the case that new, scientifically substantiated methods are not properly used in planning specifically as a result of the reluctance, and even inability, of some
practical personnel to study them seriously. We cannot recon-
cile ourselves to such a situation. Guilty parties must be penal-
ized for incorrect economic decisions that run counter to modern science and for lost national economic effect just as strictly
as those responsible, e.g., for gross violations of technical requirements. As L. I. Brezhnev emphasized in his report at the
Twenty-Fourth Congress of the CPSU, we must criticize not
only those who make mistakes but also those who do not use
every possibility for the development of production and who do
not display initiative and live a passive life.

An Effective Means

The new and complex problems that arise in life cannot be
resolved on the basis of stereotypes but require the elaboration of new scientific approaches. This is fully applicable to

the task posed by the Party: the task of securing the comprehensive acceleration of technical progress. Of the greatest importance here is the general creation of an economically favorable climate for technical progress and conditions under which economic managers could firmly count on recognition of justified additional expenditures on the creation or adoption of new equipment and on their initiative and effort being properly evaluated. At the same time, it is essential to consider in a comprehensive manner specific expenditures associated with the creation of new products (research work, mastery of production, acquisition of specialized equipment, temporary curtailment of production, etc.) and the effect that will be produced by the new product.

The effect derived from the adoption of new equipment is by no means confined to the material result that is embodied in the products produced at one enterprise during the first years. No less important, and often more important, is the contribution that is made to technical progress and to the technical potential of the nation due to this adoption, the possibility for the further production of a given product on a large scale at the same or another enterprise, and the lowering of cost to the customer. For this reason the methods of calculating the effectiveness of measures relating to new equipment should be constructed in a basically different way from the methods for determining the effectiveness of conventional economic measures. In precisely the same way the financing of measures pertaining to new equipment cannot be confined to production development funds, credits, and other conventional sources. In our opinion it should be supported from large special centralized branch and state funds significantly larger than the present new technology fund; moreover, not only during the period of adoption but also sometimes during the first years of diffusion of the new product.

In addition to everything else, such a financing procedure would make it possible to abandon the concept of higher prices on new equipment. It is hardly expedient to use the high price to cover at the expense of the customer all expenditures associated with the introduction of a new product. This can lead to an incorrect orientation in its application and can make new equipment unpopular or impede its dissemination. It would be more correct for the customer to pay for the product on the basis of an economically normal, long-term price and for the first producer to receive a higher (accounting) price from centralized funds, thereby making the production of the new item sufficiently profitable. Some economists believe that there is no justification for creating such "hothouse" conditions for new equipment; however, their arguments cannot be considered indisputable. In any case it is clear that problems pertaining to the financing of work on the creation of new equipment, to the establishment of prices on new equipment, and to the elaboration of scientific methods for determining its effectiveness must be resolved in a comprehensive manner, with due regard to the recommendations of economic science and, in particular, with the use of economic-mathematical methods.

Economic-mathematical planning methods correspond to the nature of the socialist economy and are an effective means of resolving its problems. Of course, despite the great progress that has been made in the elaboration of economic-mathematical methods, we are still far from meeting all practical demands. There is nothing surprising about this. For example, physics and mechanics have broadly applied mathematics for 200 years and nonetheless still have many unresolved problems. But as regards mathematical methods in economics, they are essentially still in the stage of formation and confirmation. It is essential to expand and improve the training of cadres mastering economic-mathematical methods, to deepen and expand research in this area, and at the same time to broadly incorporate research findings in economic practice and to enrich them with experimental testing.

The intricate nature of economic problems does not permit a formal approach to their solution and requires the use of various methodological means. We cannot hope for some universal "master key," e.g., one mathematical model, to fully resolve all problems involved in the control of a complex economic system. The pursuit of fads is impermissible here.

Apply the Achievements of Economics

Mathematical optimization methods in planning must be skill-fully combined not only with other new means (goal-oriented programs, simulation methods, etc.) but also with the traditional means of economic analysis that in a number of instances permit a more concrete and precise accounting of the specific features of one or another branch or of individual economic objects.

The correct combination of various methods in economic work, their interaction and interpenetration will best promote the improvement of the planning mechanism in socialist management, the most complete utilization of economic laws, and further successes in communist construction.

3

Optimal Planning: Unresolved Problems*

L. V. KANTOROVICH

Thirty-five years ago, Leonid Vital'evich Kantorovich's
book Mathematical Methods in the Organization and Planning of Production (Leningrad State University Press, 1939)
was published and opened up the new era in the use of mathematical methods for the organization and planning of production.

The editors of the journal requested Leonid Vital'evich
to describe the present state of this field and the areas of
the economy in which mathematical methods are still not
sufficiently applied.

In the last decade the methods of optimal economic planning
have undergone profound theoretical elaboration and serious
practical testing. They are in particularly broad use in long-
term branch planning, where optimal plans that frequently yield
a saving of many millions of rubles are compiled for scores of
important branches.

But nonetheless, at the present time we are far from making
complete use in economic practice of the enormous potential of

*Ekonomika i organizatsiia promyshlennogo proizvodstva,
1974, no. 5.

these methods. In part this is due to their novelty, to the lack of preparedness of cadres, and to other objective factors.

In the national economy there are many areas in which the application of the methods of optimal planning is particularly important.

At the same time that long-term optimal planning calculations are made for scores of branches, they are used only in rare instances in current planning. Yet the application of optimization methods for the more effective distribution of the production program among the enterprises of a branch or subbranch may yield an appreciable and immediate effect. Evidence of this point is work conducted by Gossnab and the Ministry of Ferrous Metallurgy on the rational utilization of pipe and rolling mills. An even greater effect can be expected in branches with a less intensive utilization of equipment.

Of course, the application of mathematical methods in current planning has its own difficulties: the high degree of responsibility, rigid deadlines for obtaining solutions, the receipt of information in good time, the necessity for the correction of plans, and interdepartmental coordination. At the same time, the performance of this work, in addition to the immediate material effect, will promote the improvement and recognition of optimal planning methods and the mastery of these methods by a broad range of production personnel.

* * *

Theoretical conclusions concerning the structure of economic indicators are of great practical importance. However, they have not yet won an acknowledged place and are not properly used in our planning and economic organs. The important principle of payment for capital and taking into account the capital intensiveness in price, which figured in the decisions of the September (1965) Plenum of the Central Committee of the Communist Party of the Soviet Union and which was preceded by a major debate, was only partially realized in the price revision of 1967.

As a rule the capital intensiveness of a concrete product is not taken into account, and this leads to incorrect price relationships.

Practically speaking, to date there has been no systematic accounting of rent in price, which holds particular importance for the mining industry and agriculture. The methodology of optimal planning offers sufficiently effective and objective methods for establishing prices and rents. Changes in economic indicators of agricultural production and in the interrelations between agricultural enterprises and the government on the basis of these prices and rents are a logical development of well-known decisions of the Central Committee on Agriculture. This could significantly strengthen the economy of agricultural production and promote its further intensification and specialization, the development of interfarm relations, the equalization of economic conditions for agricultural enterprises in different natural zones, and the observance of the principle of equal pay for equal work. All this could be tested on the basis of the experience of one or two republics or one or two regions in the RSFSR.

* * *

Accounting for the time factor in economic calculations is of particular importance. Even though it has received a certain amount of recognition and the methods for calculating the effectiveness of capital investments envisage the discounting of expenditures incurred at different times, nonetheless this principle is not consistently implemented in economic practice. For example, potential capital investments with an effectiveness several times greater than normative effectiveness are frequently not realized for years and years.

The accounting of intertemporal differences in the calculation of expenditures on construction and in assessing the work of construction organizations would greatly reduce the scattering of resources over numerous projects, construction delays and gestation periods, and the freezing of resources in incomplete construction.

* * *

18

While the need to calculate transportation costs according to incremental (differential) expenditures is considered to some degree in methods used in the siting of branches, this necessity is in no way reflected in rail transport rates. The systematic and correct calculation and correct payment of transport services would permit the more successful solution of problems pertaining to specialization and concentration, would eliminate the unnecessary duplication of enterprises, would permit the use of the best sources of raw materials, and would significantly reduce industry's operational and capital expenditures. This would permit correct solution of problems in the financing of transport development and in establishing the magnitude of demand for transport.

Rates also exert a great influence on improving the work of urban transit; Pravda has repeatedly written on this score.

The correct relationships and differentiation of retail prices and prices on services are of great importance. Through this alone, sometimes even without material expenditures, the living conditions of the population can be appreciably improved. However, there is no serious theory of retail prices, and practice operates eclectically.

* * *

Where technical progress is concerned, it is very important to create an economically favorable climate in which managers do not shun new technology, in which they can count on compensation for justified expenditures on measures pertaining to the realization of new technology, and in which the implementation of these measures is positively assessed. In order to make a correct economic assessment of these measures, it is also necessary to consider the major specific expenditures associated with the creation of new products (research, mastering of production, acquisition of equipment, curtailment of production) and the effect that they provide.

The effect derived from the adoption of new technology is by no means confined to the material result embodied in a product produced at the first enterprise during the first years. No less important, and frequently more important, is the contribution made to technical progress and to the technical potential of the nation by such adoption: the possibility for the further production of the same product on a larger scale at the same or another enterprise, the mastering of new production processes, the lowering of the level of expenditures, the improvement of design, not infrequently the creation of the production of other similar types of products, etc. Therefore, the methods for determining the effectiveness of measures relating to new technology must be constructed in a basically different way from the methods used in calculating the effectiveness of conventional economic measures. Nor can the financing of measures relating to new technology be limited to production development funds, credits, and other conventional sources. It must be supported by large centralized branch and national funds that are considerably larger than the present new technology fund, not only during the period of adoption but sometimes during the first years of diffusion of the new product as well.

In a number of instances it is not expedient to use higher prices to cover at the expense of the customer all expenditures associated with the adoption of a new product. This may impede the recognition and diffusion of the product. It would be more correct for the customer to pay an economically normal long-range price for the product and for the first producer to receive a higher (accounting) price for it. Then the production of the new product will be sufficiently advantageous. Nor should we fail to consider the possibility of risk. There are those who believe that the creation of such "hothouse" conditions for new technology would be incorrect. In reality it is evidently impossible to attain sufficiently high rates of diffusion of new types of products in any other way. This is confirmed by research and practice.

It is very important that the specific features of these problems (the financing of new equipment, the formation of prices on new equipment, methods of determing its effectiveness) be considered in the process of resolving them.

*　　*　　*

20

In the last ten years the new methods — mathematical model-building and optimization — have become firmly entrenched in economic science and have substantially enriched economic theory and practice. The scientific research that has been done and the experience that has been amassed have strengthened our belief that these methods correspond to the nature of the socialist economy, are characteristic of it, and therefore are particularly effective means for resolving the problems that face it. This is also confirmed by the experience of other socialist countries that are making wide use of the attainments of Soviet economic science. A number of key Party and government documents point out the positive significance that work in this area holds for the national economy.

At the present time mathematical methods are becoming an instrument for the solution of the most important national economic, social, and other problems, which increases to an extraordinary degree the responsibility of those who elaborate and apply them.

Despite the successes that have been attained, we have certainly not met all practical demands. Economic matter is extremely complex and the process of comprehension is probably infinite (it is worth noting that physics and mechanics have been applying mathematics for 200 years and are also unable to answer all their questions). It is our task to deepen, broaden, and accelerate this research, and even — without waiting until it is completed — to incorporate it broadly in practice and to perfect it in the process of its application.

The complexity and diversity of the problems require that the basic methods be supplemented by other new methods. Among them are goal-oriented programs, simulation systems, systems analysis, and expert assessments. We must, however, avoid the pursuit of fads and enthusiasm for first one method and then another.

The immense problems that have to be resolved when the new methods are introduced in all branches of the economy at all levels require the participation of numerous and specially trained economic cadres. The institutes that have specially developed the new methods (Central Economic-Mathematical

21

Institute, Institute of Economics and Organization of Industrial Production, Institute of World Economics and International Relations, scientific-research institutes of USSR Gosplan and RSFSR Gosplan) are not equal to this task. It is essential to train such cadres in all economic institutes, and the methods themselves must be adopted by our planning and economic organs.

It is essential that all economic cadres receive the appropriate training, that mathematical cadres oriented toward working in economics be trained.

We should think of changing the programs for the economic training of engineers, since all of them need a greater degree of sophistication in economics.

Finally, the complexity and importance of the subject of economic science require critical analysis and public discussion of the new principles and methods. There have been repeated critical discussions and debates (on the criteria of optimality in branch planning, on normative effectiveness, etc.). There must be more critical discussion of complex theoretical questions, concrete problems, and practical methods. It would be a good idea to conduct such discussions on the pages of the Economic Gazette [Ekonomicheskaia gazeta]. In discussing the usefulness of criticism I refer to practical, positive criticism based on the careful study of the criticized works and aimed at ascertaining scientific truth. There is also another type of criticism that only impedes the work, and it is by no means harmless.

Advocates of optimal planning have been repeatedly reproached with the fact that the application of these methods, which have been developed and propagandized for a long time, is still quite limited. It should be said that in addition to basic difficulties, some shortcomings in the organization of the introduction process, and the inertia of certain economic organs, no small part has been played by the unfounded criticism to which this research has been subjected (in particular by Ia. Kronrod and A. Kats). The practical application of evaluations of the optimal plan and of those economic indicators that were the subject of the greatest attacks has been particularly retarded.

The enormous scale and profound interrelationships of modern socialist production make the tasks of economic management particularly complex and important. They are growing more complex thanks to the continuously increasing rate of technical progress and the need to take new ecological, social and political problems into account. On the other hand, the significant progress that has been made in the development of economic science and its enrichment by new means are encouraging and important.

4

A Dynamic Model of Optimum Planning*

L. V. KANTOROVICH

A sharp increase in the scope and in the complexity of the production relationships of socialist construction points insistently to the problem of a radical improvement in the methods of planning and economic calculation.

It is stated in the Program of the CPSU that in a communist society "the highest stage of a planned organization of the entire social economy is reached, and the most effective and rational utilization of material wealth and labor resources is assured for the satisfaction of the growing requirements of the members of society." (1)

In other words, an optimal plan with regard to economic activities will be carried out under communism. Accordingly, the problems of constructing a communist society require even now a gradual transition to optimal planning and to optimal solutions in all our economic activities.

It is for this reason that the problems of optimal planning have been attracting the attention of Soviet economists in recent years, especially those economists who are concerned with the application of mathematical methods in economics,

*Planirovanie i ekonomiko-matematicheskie metody, Moscow, "Nauka" Publishers, 1964.

without which it is impossible to envisage any effective work with regard to these problems. One cannot fail to note, in this connection, that the optimum aspects of planning occupy an important place in the works of V. S. Nemchinov, particularly in his book Economic-Mathematical Methods and Models. (2)

The study of optimal problems in the economy requires the special mathematical methods of linear and nonlinear and dynamic programming which have been created as a result of the work of Soviet and foreign scientists and which have been successfully applied so far to the economic problems of individual enterprises and sectors.

In our own study (3) the problems were formulated that relate to the wide utilization of these methods in the planned direction of a socialist economy as a whole and to the construction of its economic indicators. Two mathematical models of optimal planning were considered there: (1) a static model that simulates current planning, and (2) a dynamic model that simulates long-range planning. It was indicated, however, that it is the dynamic model which should be regarded as the basic one, while the static model is more specific and subordinate to it.

Because of methodological considerations, however, it was the static model that was considered first and in greater detail, and in a certain measure this has made it more difficult for the reader to grasp the overall conception. It is for this reason that the dynamic model, and the analysis and conclusions associated with it, have attracted less attention. Yet that model reflects more fully the actual tasks of planning. It can be effectively employed for planning both sectors and the national economy as a whole and can serve as one of the instruments for studying the quantitative regularities of the process of expanded socialist production.

The aim of the present study is to shed light more clearly on the structure of the dynamic model and on the ways in which it may be implemented and to draw certain conclusions to which one is brought by analysis of this model.

1. Description of the Dynamic Model
of Optimal Planning

We consider that the economic indicators of a socialist economy are inseparably linked with the rational organization of production and of the process of expanded reproduction, proceeding on the basis of a scientifically constructed optimal plan. Accordingly, a model of the optimum economic plan will constitute the foundation of our analysis. Let us consider first a linear dynamic model, but let us also indicate at the same time the possible ways for making it more accurate and more general.

Basic ingredients. We distinguish in this model the following ingredients: different types of raw materials, of production and of services, taken in an aggregated form (grain, fuel, shipments), in terms of a variety of physical units (tons, meters, items), in conventional physical units (equivalent tons of a conventional fuel, tons-kilometers, 15 h.p. tractors, conventional fodder units), or else in conventional value form (constant prices). At the same time, the less transportable types of production (ore, fuel, cement, electrical power) are localized: they are connected via the index i with a particular region. In all cases products of a particular type are differentiated with respect to the time period during which they are produced. It is assumed that the overall planning period is divided into a series of time intervals (years, quarters) $t = 1, 2, \ldots, T$. In this way the different types of ingredients are characterized by two indices: i and t e.g., $G_{i, t}$.

Types of labor. Labor is differentiated by profession and qualification, and possibly also on the basis of sex and age characteristics as well as by region and time period of its utilization.

Production capacities are differentiated by types, sectors, etc., by time period at the time of their introduction and intensity of utilization, and also by location. They may refer to entire enterprises, areas of production, individual pieces of equipment, highways, etc.

Natural resources. Land, forests, water resources, mineral deposits are also differentiated in accordance with quality and

26

location and are measured in corresponding units.

Production (technological) methods. The methods of production of particular types of products or groups of products are characterized by the necessary inputs of different types of ingredients and the resulting output. Among them one may note specifically those methods of production that are completed within a single time interval. In such cases all the input and output elements refer to the same interval. Accordingly, such a method is described by the vector

$$a_1^{(s)}, \ldots, a_n^{(s)},$$

where $a_i^{(s)}$ denotes, depending on its sign, the input or output of the ingredient $G_{i,t}$ corresponding to a unit application of the s-th method. In fact, the components $a_i^{(s)}$ themselves may depend on

$$t : a_i^{(s)} = a_i^{(s)}(t).$$

This makes it possible to characterize technological progress, for instance, by a reduction of the normative inputs. In addition a given s-th method may be applied only after a certain $t \geqslant t_s$.

In particular, these methods include modes of transportation that bring about the displacement of products from one region to another with definite inputs and through the utilization of different transportation possibilities.

There are methods that affect several periods, in which ingredients relating to different periods operate at the same time. A characteristic example of this kind of method is the production of means of production, in which a production capacity (a machine or an enterprise) is produced in the course of one or two periods that can subsequently be used over several years.

Another method of this type is the storage of a certain product over a given period.

Still another is provided by the application of a production method over a single period, if one takes into account the

27

changes in the production capacities that result from its utilization (aging, wear, etc.). The education and training of labor and its transfer from one region to another should also be regarded as such a method.

Finally, the bringing into use of natural resources comes under this category as well.

Resources. It is necessary to characterize the initial resources and, first of all, the initially available stocks, production capacities, natural resources under active exploitation, available manpower, and enterprises under construction. Next, for the entire period under study the natural resources and the possible prospects for their expansion should be described, as well as the constraints and requirements relating to their utilization. The manpower resources are indicated, for example, through forecasts of population growth that include demographic descriptions of its composition, and also through indications of the share of the population that is engaged in productive work and of the share of labor that can be applied in the production and nonproduction spheres. This may be done separately for the rural and the urban population.

Consumption. The basic characteristics of the composition of the final products necessary for personal and social consumption must be indicated, and its growth over the years of the entire planning period under review must also be tentatively described. (The rate of growth may be indicated more accurately in the course of the process of solution.) The allowable limits on the variation of the composition of final output and the equivalencies for substituting particular types of consumer goods by others, etc., may be indicated. Consumption may include not only consumer goods proper but also those means of production that are required in the nonproduction sectors of society (e.g., science, defense) as well as the production capacities that they require. Deliveries for personal consumption may be given independently of the way in which labor is used or else in accordance with its distribution by professions and sectors (for instance, by taking into account the salary levels).

Optimality criteria. The optimality of a plan may be

characterized in different ways. For example, that plan may be regarded as optimal which brings about, with given resources and the satisfaction of given volumes of consumption, either the attaining of a maximal productivity in the last year, or else the attaining in the shortest time period of given norms and composition of consumption, or even the attaining of given rates of growth and of consumption with minimal expenditures of labor (for instance, together with a reduction of the length of the working day). The plan that is optimal with regard to any of these conditions will be an <u>optimum production plan</u> in the sense that there will not exist another plan that could provide equal levels of output over all the periods with the same resources as well as equal resulting production capacities (and for individual types of production, even larger ones).

Let us write the above conditions in mathematical form. For the sake of definiteness let us divide the ingredients into four groups:

G_i $(i = 1, 2, \ldots, n_1)$ are the primary resources (resources of productive labor and natural resources);

G_i $(i = n_1 + 1, \ldots, n_2)$ are the factors of production (categories of labor, production capacities, actively exploited natural resources);

G_i $(i = n_2 + 1, \ldots, n_3)$ are the intermediate products; and

G_i $(i = n_3 + 1, \ldots, n_4)$ are the final products.

Let $x_{i,t}$ denote the balance (net sum) for the i-th ingredient over the period t. Then the primary resources over all the periods are given by the inequalities

$$x_{i,t} \geqslant - L_{i,t} \; (i = 1, 2, \ldots, n_1; \, t = 1, 2, \ldots, T), \qquad (1)$$

where $L_{i,t}$ expresses the availability of the i-th resource in the interval t.

The production capacities and other factors in the initial period are given by

$$x_{i,1} \geqslant - L_{i,1} \; (i = n_1 + 1, \ldots, n_2). \qquad (2)$$

29

For the production factors, and for the intermediate and final products, the following inequality must hold:

$$x_{i,t} \geqslant 0 \ (i = n_2 + 1, \ldots, n_4; \quad t = 1, 2, \ldots, T). \qquad (3)$$

The dimensions of consumption are assumed to be given for every year

$$C_{i,t} \ (i = n_3 + 1, \ldots, n_4; \ t = 1, 2, \ldots, T),$$

As a matter of fact, various modifications of this condition and of the way in which it is taken into account are possible.

Production methods. Each (s-th) method is generally characterized by the matrix

$$\|a_{i,t}^{(s)}\| \, (i = 1, 2, \ldots, n_4; \ t = 1, 2, \ldots, T), \ (s = 1, 2, \ldots, S).$$

The plan. The plan is characterized by indicating the intensities of the application of the methods r_s ($s = 1, 2, \ldots, S$). The balances of the plan with regard to the individual ingredients are determined by the formula

$$x_{i,t} = \sum_{s=1}^{S} r_s a_{i,t}^{(s)}.$$

The consumption is indicated in the form of a method of production

$$(0, \ldots, 0, - C_{n_3+1, \, t}, \ldots, - C_{n_4, \, t}, 1_t) \quad (t = 1, 2, \ldots, T)$$

through the introduction of a new ingredient

$$G_{n_4+1, t}$$

that characterizes the process of consumption (a set of consumption products).

The bound

$$x_{n_4+1, \, t} \geqslant 1, \qquad (4)$$

is imposed upon it, which requires the application of the above method with an intensity $\geqslant 1$.

A different notation is required for more complex descriptions of the consumption conditions. For example, the provision for substitution in consumption of one group of products by others may be denoted by including several methods of obtaining the ingredient

$$G_{n_4+1,\ t}\ ,$$

symbolizing an assortment of consumers goods.

Finally, the optimality condition, too, can be written in different forms: for instance, the maximization of the final output in the final year; the maximum rate of growth of final output; the same, in the presence of specific requirements concerning the level of production capacities in the final period of the plan; or else a given rate of growth of final product, providing that it is attained with minimal labor expenditures (i.e., with a maximal reduction of the length of the working day). All these conditions are also written in the form of a system of inequalities and by specifying a function that reaches an extremal value.

For example, a maximal rate of growth in the satisfaction of social and personal needs leads to the requirement that the maximum α be found for which the problem is solvable, under the conditions:

$$x_{n_4+1,\ t} \geqslant (1 + \alpha)^t \ (t = 1, 2, \ldots, T), \tag{5}$$

i.e., that for the given composition of the consumption fund the initial rates of growth of production in each year be exceeded as much as possible. The conditions corresponding to the production capacities at the terminal period are written as the inequalities of the form

$$x_{n_1+1,\ T} \geqslant x^0{}_{n_1+1}$$

$$\cdot\ \cdot\ \cdot\ \cdot\ \cdot\ \cdot\ \cdot\ \cdot\ \cdot\ \cdot$$

$$x_{n_2,\ T} \geqslant x^0{}_{n_2},$$

31

where x^0_i is the given minimum volume of production capacities of the i-th type at the moment T.

Other forms of the optimization conditions are also possible.

In any case, with any of the above or other possible selections of a criterion of optimality the resulting plan will be an optimal production plan in the sense that there exists no other plan which would ensure the same output for consumption and the same basic real funds at the end of the period with a smaller expenditure of resources (or, inversely, a larger output, with the same expenditure of resources). This alone is sufficient for our further conclusions.

From the economic point of view the dynamic model for long-term planning differs in a fundamental way from the current planning model that is described, for instance, by the model of the basic production problem. (3) From the mathematical point of view, however, it formally corresponds entirely to the scheme of that model, with the only difference that the dynamic model contains many more ingredients. Accordingly, the theory relating to the characteristics of an optimal plan applies to it as well. On the basis of this theory as well as of other theories of linear programming, we obtain the following theorem.

2. A Theorem Concerning the Characteristics of an Optimal Plan

An optimum production plan, and in particular a plan described by conditions (1)-(4), is characterized by the presence of a dynamic set of objectively conditioned valuations corresponding to all ingredients at all moments of time

$$\xi_{i,t}$$

(i.e., solutions to the dual problem) such that the following conditions for them are fulfilled:

1) all the methods applied in the plan are justified (profitable) in terms of these valuations

$$\sum_{i,t} a_{i,t}^{(s)} \, \xi_{i,t} = 0, \ \text{если} \ r_s \neq 0;$$

2) all feasible technological methods are not more than justified

$$\sum a_{i,t}^{(s)} \xi_{i,t} \leqslant 0 \quad (s = 1,2,\ldots, S);$$

3) $\xi_{i,t} \geqslant 0$;

4) if any inequality occurs in the constraints for a particular type of resource or output, the corresponding

$$\xi_{i,\,t} = 0.$$

If the optimal plan is a production plan, then such valuations exist. Conversely, the existence of such valuations indicates that the plan is an optimum production plan. More specifically, if the plan is feasible, the constraints (1)-(4) are fulfilled, and there exist valuations satisfying the conditions (1)-(4), then the plan represents an optimal solution of the optimal programming problem.

The fact that the long-run planning problem is described mathematically by a linear programming scheme makes it possible to solve it with the aid of the usual universal methods of linear programming, such as the method of successive improvements in the plan, the method of correcting multipliers, and the simplex method, even though it is true, of course, that the considerably larger number of ingredients in this problem makes its solution more complicated. It is by no means a hopeless task, nevertheless, if electronic computers and modern methods are employed, particularly since the specific character of the matrices of these problems makes possible the application to them of special procedures. Without discussing this question in greater detail in the present article, I shall merely express my conviction that the mathematical difficulties will be successfully overcome in connection with the formulation of real problems, including those that will be encountered in the long-run planning of the economy.

Let us consider a fundamental economic conclusion, which follows from the above theorem relating to long-term planning. This is the fact that an optimal plan is characterized by the presence for all ingredients of a dynamic set of valuations conditioned by the optimal plan itself and by the conditions that give rise to it (resources, technology, the direction of the development of the economy, the criterion of optimality). These valuations represent relationships of equivalence for different ingredients with respect to their effectiveness under the conditions of an optimal plan. In particular, the presence of these estimates thus provides a means for reducing to a single equivalent not only different qualities of production and of inputs, but their occurrence in different time periods. The economic significance of these valuations of production lies in the fact that they characterize the relations to each other of the labor expenditures that are necessary for producing a unit of output of one or another type. This refers, moreover, to differential expenditures, that is, expenditures per unit output associated with an increase or, correspondingly, a decrease in the output of a given production.

The reduction of inputs relating to different time periods to a single equivalent through the use of a dynamic system of valuations becomes especially apparent in its economic aspect if a generalized index is established on this basis, such as the norm of effectiveness of capital investments.

After selecting a certain standard assortment of products and of production factors

$$(\bar{\alpha}_1, \ldots, \bar{\alpha}_{n_i})$$

let us establish a scale of prices such that the price of this assortment remains constant. Let us introduce the multipliers λ_t from the condition

$$\sum_i \bar{\alpha}_i \xi_{i,t} = \lambda_t \sum_i \bar{\alpha}_i \xi_{i,1}.$$

A Dynamic Model

The quantities

$$\left(\frac{\lambda_t}{\lambda_{t+1}} - 1 \right)$$

will then characterize the norm of effectiveness in each period of capital investments, while the ratio

$$\frac{\lambda_{t+1}}{\lambda_t}$$

indicates the coefficient which relates a given year's expenditures and outputs to those of the preceding year.

If, however, we introduce another system of dynamic valuations

$$\xi'_{i,t} = \frac{1}{\lambda_t} \xi_{i,t},$$

then these valuations will correspond to a single scale, and their alteration in time will represent a relative increase or decrease of the valuation of the given ingredient by comparison with others. Let us note that the assortment of products forming the basis may vary rather than be fixed, and the reduction is carried out by the so-called chain method.

3. Analysis of the Dynamic Model

In drawing a comparison of the dynamic model with the static model of current planning, one cannot fail to note a number of specific characteristics and advantages of the former in relation to the latter, as a result of which the dynamic model is seen to be much less arbitrary.

An important shortcoming of the static model is that in current planning the utilization of available production facilities largely predetermines the technology and organization of production and limits severely the possibility of employing other variants of the production process. This reduces substantially the possibilities for using linear programming and its effectiveness. In the dynamic model, on the other hand, insofar as the creation of the necessary production capacities is also included in it, these difficulties are removed, and one may take into account in the analysis the entire range of realistic variants of the development of production that are provided by contemporary technology.

In the model of current planning a large measure of arbitrariness was introduced by the need to indicate the output targets for all final outputs, which included not only consumer goods but also means of production. On the one hand, such a specification limited very much the variation of the production plan; on the other, it was the source of much difficulty in regard to determining the targets relating to the production of means of production, since that target could be determined realistically only by referring to a plan for the subsequent development of production. In fact, however, it is precisely in relation to its correct selection that the advantages of an optimal plan are largely to be found. In a dynamic model only targets relating to consumer goods are required, and even then in a quite general form, while the selection of a program relating to means of production occurs in the course of the construction of the plan itself.

Finally, in the presence of the conditions that relate to the current planning model there may occur a significant measure of instability in the valuations, in their variation as a result of accidental circumstances as well as zero values, and, generally, a significant deflection of valuations from their natural stable values. In the dynamic model these shortcomings, too, are considerably reduced. If a certain product is produced in excessive quantities, it will not have, as a rule, a zero valuation, since in a dynamic context one may employ the "method of storage" with a utilization of the given product in a subsequent period.

In an optimal dynamic plan disproportions are eliminated altogether, as a rule, or else reduced.

A Dynamic Model

At the same time, the fact that objectively justified dispro-portions associated with unforeseen technological progress or newly created requirements are reflected in the valuations, in-cluding those for means of production, represents an advantage. In this respect we cannot agree with V. Novozhilov. (4)

It is necessary to state that the shortcomings of the current planning model that have been mentioned arise if it is treated as an independent model. If the current planning model is viewed (as it should be) as a component or segment of a dy-namic plan, corresponding to a single period, with resources and targets coordinated with this dynamic plan, then it, too, will be free of these shortcomings.

Another aspect that should be noted concerns the form in which the specifications relating to consumption are expressed. The composition of that production which serves social and es-pecially personal needs, and its rational proportions, depends in a large measure on noneconomic factors. Yet the fact that we assume this composition to have already been selected, if only in an aggregated form (or else assume the other forms of determining it mentioned above), is not very essential. Even if this composition is selected not in the best way, and it is ex-pected that it will be made more specific and improved in the future, such a corrected composition can also be produced with approximately the same production capacities. In any event, such a correction will not change the essential elements of the solutions of the plan for the development of the main sectors of Group A.

The forms of the optimality criterion that we have accepted may also be questioned. This may be done, specifically, be-cause an appraisal of the quality of a plan depends to a certain extent on an appraisal of the potentialities of the economy with regard to growth in the subsequent period. One of the methods for making the selected criterion more specific is its a poste-riori reformulation on the basis of the results of a solution and a subsequent recalculation of the plan. For example, in apprais-ing a plan, an appraisal of production capacities at the end of the planning period is important. It can be expressed more adequately by utilizing neither physical indices nor conventional value indices based on constant prices, but rather estimates of production capacities determined by the optimal plan itself and also the dynamic valuations of production and the value of the norm of effectiveness obtained in the calculation of the plan.

One should state, incidentally, that the choice of one or another reasonable criterion of optimality is not of such decisive significance. The choice of a criterion does affect substantially, of course, the plan as a whole, but it does not affect very perceptibly the solutions and indices of the initial years of the plan which are especially important for us. Indeed, if one has in mind the principle of uninterruptedness in planning, one should expect that those parts of a long-term plan that relate to its final intervals will always be subsequently revised in the light of new data and of new targets.

The two problems noted above are closely related to the so-called problem concerning the optimal relation between consumption and accumulation. With regard to this problem, in addition to economic questions, noneconomic questions (the political situation, social factors, etc.) are also significant. In other words even when employing mathematical-economic methods, it is not possible to establish this relation in final form even in principle, on the basis of purely economic considerations alone. In the course of the construction of an optimal plan, however, one may determine fairly narrow boundaries for this relation. On the one hand, the upper boundary is determined by the feasible rates of growth of consumption that could be achieved. The lower level is a certain minimal necessary level of consumption. Finally, one may seek to determine the boundary for consumption on the basis of the dependence of the productivity of labor on real wages.

There does remain, nevertheless, a certain area of indeterminateness for this relation. In selecting it one may then employ calculations of the optimal plan based on several variants of this relation, together with a subsequent comparison of the results attained through these variants from the point of view of the entire set of economic indices and a simultaneous consideration of noneconomic factors as well.

Let us consider next the role of the assumption of linearity which is present in the dynamic model, the extent to which it is justified, the significance of this assumption, and the possibilities for dispensing with it, as well as the effects of taking nonlinearity into account.

It is necessary to note, first of all, that the hypothesis of linearity is much more justified than it might at first appear. Indeed, in an individual enterprise there may exist a fairly strong dependence of inputs per unit of production on the level of output, i.e., a nonlinear dependence. Within the boundaries of the overall economy, however, in the context of long-term planning, and when normal (close to optimal) sizes of enterprises exist, an increase in output is usually associated with a corresponding increase in the number of enterprises; and in such cases proportionality (especially with regard to processes of mass production) is fairly accurately maintained. It should also be noted that despite its linear character the model can reflect effectively several nonlinear relationships as well. Thus a nonproportional increase of expenditures because of the utilization of less effective sources (a convex nonlinearity) may be represented by introducing several production methods with different volumes of application instead of merely one. A nonlinear dependence of the level of output on the rate of growth can be expressed by a chain utilization of several production methods, written for a number of years.

There are cases, however, when nonlinearity is an essential element and possesses, moreover, a nonconvex character. In particular, this applies to the case when a product is produced by very few enterprises (one enterprise or two) while both capital and current expenditures per unit output decrease substantially as the volume of output increases (serial production of machines or instruments, railway shipments in a given area, etc.). In such a case the linear programming scheme and the computing techniques must be enriched by the methods of nonlinear and integer (and sometimes dynamic) programming, for instance, by always choosing integer values for the intensities

of some of the methods in the model described above.
Even when this is taken into account, however, the fundamental economic conclusion concerning the presence of a dynamic system of valuations for the optimal plan remains essentially valid. If we imagine that an optimal plan has been constructed (with due account of nonlinearities), then all possible small variations of the plan may be viewed as an application of some linear technological methods with different intensities. For certain variants the intensity may be only positive or only negative, as, for instance, when it is possible to reduce the output of a product with a given method, but not to increase it, or it is possible to do so only with other inputs. Thus, if we take such an optimal plan as the basis, its possible small variations taken together may be described by a linear programming problem; since the plan is optimal with respect to these variations, dynamic valuations for all ingredients will be connected with it, so that the fundamental propositions (1)-(4) characterizing an optimal plan will hold, with the only difference that they will be effective with respect to a situation in which these methods are applied with not too high an intensity. The economic meaning of these valuations under such conditions is that they describe differential expenditures or effects associated with the production or utilization of a unit of the corresponding ingredient.

It should be noted that apart from complicating the determination of the optimal plan, nonlinearities and integer requirements will be reflected in the system of valuations and in the corresponding conclusions concerning price formation. It has emerged that for outputs of this type the price must be constructed at the level of differential expenditures rather than the total expenditures required for their production. In this way the production of these outputs, with due account of expenditures for the establishment and bringing into operation of capital investments, need not necessarily assure profitability in terms of the cost calculations (i.e., the theorem concerning the characteristics of an optimal plan is not always met). In other words, there exist cases when the requirements of cost accounting need not be observed in dynamic planning. In reality the organization of such types of production turns out to be profitable just the same, if its full effect on the economy is taken into account. For example, low railroad rates established in accordance with differential expenditures will not make it possible to attain the level of profitability that will assure the normative effectiveness for the expenditures on the construction of the given railroad, but its profitability in accounting terms will be attained if one takes into account the reduction of expenditures in sectors of production that utilize the given railroad, i.e., if one takes into account the full economywide effect. For the same reason the prices of such products must also be accounted differently both in current plans and in the analysis of the effectiveness of capital investments.

Definite conclusions follow from a consideration of the stochastic character of the problem.

The normative expenditures of different methods, especially in the case of forecasts for future years, the data on resources, and, in particular, on natural resources, and the calculated requirements and demand in future years are actually stochastic values, which are known to us only with a certain probability. For this reason the problem of constructing an optimal plan must also be regarded as a problem in stochastic programming. Apart from complicating the process of finding solutions, this circumstance again affects in a qualitative manner the appraisal of the effectiveness of decisions and of price formation. Because of this circumstance a certain preference should be given to the key sectors and to the universal products by comparison with the specialized sectors, since in the latter the uncertainty of data and the dependence on demand are larger. This preference should be expressed in the admission in the first of these sectors of investments with a somewhat lower level of effectiveness than in the latter. In the presence of several nearly equivalent solutions a mixed strategy is preferable, in which different production methods are combined. For this reason the investigation of the stochastic formulation of the dynamic model is of real interest.

Finally, an essential element is provided by an extension of the model with regard to the utilization of labor. This can be carried out within the framework of the linear model.

In the model described above a division of labor in accordance with types and professions was considered and its resources were taken into account in a corresponding way. The size of consumption was not connected with specific directions and forms of utilization of labor. In other words, consumption was considered independently of the character of utilization of labor, and it was assumed that the problems of distribution — wages and retail prices — were to be solved separately. Actually, however, the same kind of labor can be utilized with different intensities in different situations, depending on the working conditions in the sector under study, the remoteness of enterprises, and the wage-rate system (payment by the hour, by the piece, in accordance with progressive norms, etc.). Moreover, a mutual interconnection exists in some cases between the productivity of labor and the form and level of its remuneration. The corresponding changes should then be introduced in the model. In addition to the physical inputs of labor related to the corresponding production methods, expressed in man-days and determining the volume of the required manpower, the model must also include wage payments — for instance, in the form of deviations from averages. In accordance with this the constraints must only include the number of workers of different types. At the same time, the total volume of consumers goods should be related to wages in more general terms, without taking the composition of the consumption fund into account, or else divided into income groups with corresponding adjustments in the composition of the consumption pattern. As for that part of the social consumption fund which provides security to workers, it should rather be related to the number of workers and the place of employment and, to a smaller extent, to the wages. A more accurate description of the optimal production model, however, that would take into account these circumstances would require an elaborate study of the interdependence between productivity of labor and real wages, which seems to be quite a difficult problem. Therefore, it appears rational for the time being to express these factors in the model in a simplified way.

The initial level of the consumption fund intended to meet the wage expenditures of those who are engaged in the productive sphere is calculated on the basis of certain average levels of wages and retail prices (including the tax on turnover), taking into account the dynamics of demand. Accordingly, a definite assortment of components of real wages will correspond to a given level of monetary wages. In describing the production methods it is necessary to indicate, in addition to the number of workers of particular types, the level of wages corresponding to a given organization of labor and wage system (and the resulting productivity of labor). This determines the component of the consumption fund that arises in connection with the utilization of a given production method. In this way the consumption fund calculated a priori is replaced by an amount of consumption that is determined in the process of solving for the optimal plan. Thus the production method vector must contain, together with the amount of manpower $a_i^{(s)}$ of the required kind, the category of its remuneration, $v_i^{(s)}$, i.e., the inputs for each method must include the inputs of the corresponding assortment $G_{n_4 + v_i^{(s)}}$ of consumer goods in the amount

$$a_{n_4 + v_i^{(s)}}^{(s)} = a^{(s)}{}_i,$$

if it is assumed that the assortment unit has been calculated in relation to one worker of the given wage category.

4. The Prerequisites for Applying the Dynamic Model

In considering the model that we are proposing, it is natural to note in it a definite measure of complexity, both from the point of view of the mathematical equipment that is employed and from the point of view of applying it to an economy — that is, of obtaining the normative and statistical data required for this purpose. In particular, one may note its complexity by comparison with other models, for instance, the model of Neumann, the dynamic model of Leontief, and the model of Lange.

It seems to us, on the one hand, that despite the measure of complexity that does exist in our model, it is quite applicable. On the other hand, this complexity is associated with the real features of current socialist production and hence it cannot be eliminated as a matter of principle if one seeks to make it reflect the actual economy with a sufficient degree of correctness. In particular, it is an essential characteristic of the model that it can reflect characteristic features of contemporary socialist production — such as optimality, variety of technologies and substitutability of products, the special role of labor as the only source of value, the presence of exhaustible primary factors, the presence of lags, continuous technological progress, etc.

While the Neumann model does to a certain extent take into account the first two elements, it does not consider the others and views the development of an economy in terms of a yearly cycle recurring on a constantly expanding scale. A particularly impermissible feature of that model, as well as of the dynamic models of Leontief and of Schwartz, is their neglect of labor as an altogether special and fundamental factor of production. In these models labor is simply "excluded," and it is assumed that labor may be obtained and applied in any required amounts. In other words, the actual rates of growth of the population are ignored, and a reserve of labor (unemployment) is assumed, which does not correspond at all to the principles of constructing a socialist economy. For this reason, in particular, an attempt to determine the norms of capital investment in a socialist economy on the basis of Neumann's model is completely unjustified, even as a rough approximation.

The norm of effectiveness depends, in an essential manner, of course, on the requirements for capital investment. In the models of Neumann and of Leontief this requirement is taken into account, but only in connection with the expansion of the scale of production. In reality, however, this requirement arises in connection with two other factors as well, which — in contrast to these two models — our model does take into account. More specifically, capital investments are needed for the creation of new real funds, associated with technical progress, and with new requirements, and also for the replacement of old funds by new ones in connection with their obsolescence, and in order to achieve increases in the productivity of labor so as to release it to new sectors.

In constructing a mathematical model of the development of an economy, it is important to achieve the greatest possible measure of approximation to reality. The implementation of the model presents a special problem. There the actual obtaining of the necessary parameters on the basis of statistical and normative data is vital, sometimes even at the cost of a conscious sacrificing of accuracy in the model.

Let us consider briefly one possible way of applying the dynamic model of the economy.

One of these ways is the statistical method of constructing the initial data, a method similar to the construction of a matrix of interindustry relations. The essential difference in our case consists in the following.

First of all, each sector must be divided into several groups of enterprises that differ in the technologies they apply whenever there exist several essentially different technologies or enterprises that differ sharply in their technical level (modern enterprises and older enterprises). A structure of expenditures relating each group to other sectors must be constructed.

The latent capacities of the groups of enterprises must be determined, and the constraints on the rate of development of each group that are imposed by natural conditions must be established, or else their expenditure of exhaustible natural resources must be limited. The construction costs of production capacities of given types must be calculated. The simplest approach to this is the calculation of expenditures on the basis of a determination of their real fund intensity through valuations of basic and circulating real funds and with due consideration of the share in this of construction and assembly work. The average time of construction must also be known.

It is necessary to calculate transportation expenditures for each group of enterprises (with regard to materials and finished products) and to prepare forecasts relating to technological progress by groups of enterprises — in particular, with regard to the normative expenditures for production and construction.

Finally, data must be obtained concerning the share of the population that is able to work, and also concerning final consumption.

Under these conditions the dynamic model in its very first approximation can now be constructed. If one assumes, for instance, that there are 80 sectors, the number of groups of enterprises will reach 200 to 300, and this will essentially determine the number of types of ingredients in the model. It would seem that the calculation of a model of these dimensions corresponding to a period of 10 to 15 years is quite feasible with the aid of modern electronic computers. Its calculation may yield, in a very rough approximation, a scheme of an optimal development of the economy and some of its indices. A further elaboration of the model will require that enterprises under construction be taken into account and that regional calculations for most sectors be carried out, taking into account the location of natural resources and of existing enterprises. Subsequently, more accurate models of individual sectors and regions can be introduced, and then integrated into the economywide model. Finally, 2-product or 3-product dynamic optimal models of the roughest kind, based on indices of the economy as a whole, may also be applied to a certain extent in order to obtain the approximate values of certain economywide indices.

5. Conclusions

In spite of the fact that calculations of the optimal dynamic model of the economy have not yet been carried out and will apparently not be carried out in the immediate future, it is important to emphasize that a theoretical analysis already makes it possible to arrive at a number of conclusions that may have substantial practical significance.

1. The long-term plan for the development of the economy and of individual sectors must be constructed on the basis of a model of optimal planning.

A calculation of this type, which takes into account, on the one hand, the current state of the economy and its resources, and modern technological achievements and possibilities, on the other, can make possible considerably higher rates of development, a more rapid introduction of technological progress, and a more flexible and effective technological policy than the widely used method of planning on the basis of the levels already attained.

2. In addition to a long-term optimal plan, the dynamics of the prices conditioned by the optimal plan should be determined, which reflects the dynamics of the full socially necessary expenditures on production — that is, of prices that will serve as the basis of concrete planning and technological decisions, as well as of the selection of a technological policy.

3. Together with the valuation of different types of output, a dynamic system of conventional valuations of the factors is established, which raises the productive power of labor: more effective limited natural resources, production capacities (rental valuations). Aside from their inherent significance, these valuations are a means for establishing a valuation of production on the basis of planned expenditures for production and of a consideration of conditions of production, which makes it possible to express labor inputs as averages and obtain the socially necessary expenditures. Because of these valuations of factors, and in spite of the fact that the value of production (the full socially necessary expenditures) contain not only the direct expenditures on production alone, it becomes possible to calculate full expenditures on the basis of direct expenditures.

4. Together with other factors, the dynamic system of valuations also depends on the objective relationships between the requirements for products and the possibilities for producing

them. The system of planned prices of products that is calculated in the course of the construction of an optimal plan makes it possible to regulate the rational distribution of products more flexibly and more effectively, with due account of the inevitable changes in those relations that arise in connection with changes in needs, technological progress, etc. It makes it possible to achieve this much more effectively than does the spontaneous mechanism of price formation in a capitalist economy.

In particular, this points, in practice, to the rationality of employing more dynamic and more flexible systems of planned and calculation prices that depend on the place, time, even season, etc., and that make it possible to attain the most effective use of raw materials and other materials and the distribution of finished products.

5. The taking into account of rents and of rental valuations of equipment in economic calculation, in the system of taxation, and in the prices of outputs provides an important means of attaining the most effective and most intensive utilization of the natural resources and basic real funds for the welfare of society.

6. In the course of the calculation of the optimal long-term plan, the norm of effectiveness of free capital investments is determined, which serves as the basis both for the reduction to a common measure of expenditures and results occurring in different periods and for the calculation of the effectiveness of capital investments. This norm is determined by the volume of means directed to accumulation, by the need for capital investments, and in particular, by the rates of technological progress and the level that it has attained, etc. At the present time this rate must be quite high in the USSR, apparently no less than 25 to 30 percent.

The norm of effectiveness thus established must be taken into account not only in drawing up new technological projects but also in economic calculation in order to establish the cost of production and construction costs. This would stimulate rapid construction methods and the concentration of means in accordance with projects and in time. It is possible that it may be expedient to apply it to stimulate an acceleration in the turnover of circulating real funds and thus prevent an excessive accumulation of stocks.

7. In the calculation of the optimal plan different kinds of labor also receive conditioned valuations. In this connection, in order to stimulate economizing and a correct utilization of labor, it is expedient in some cases to introduce, apart from wages, special payments by enterprises for the utilization of certain types of labor. This procedure, together with the taking into account of available labor resources in the planning of capital investments, will contribute to the fullest and most effective utilization of labor resources.

8. The introduction of the rental valuation of equipment into cost accounting, by including the corresponding payments of enterprises and by taking into account their share in the production costs and in the price of products, will contribute to a more correct evaluation and utilization of products produced by means of expensive or scarce equipment.

At the same time, the adoption under these conditions of the rate of profit as the main index in appraising the quality of the operation of an enterprise will stimulate the fullest use of equipment, an interest on the part of enterprises in increasing their output plans and also the volume of orders, and will also stimulate a rational specialization and cooperation.

9. For agricultural enterprises a particularly essential significance attaches to the introduction of rent payments in accordance with the size of the rent as it is objectively calculated on the basis of the optimal plan for the location of agricultural production. Apart from leveling out the working conditions in agricultural enterprises in different zones and also the remuneration of labor, the introduction of such payments will stimulate a rational location and specialization, and also more intensive forms of agricultural production, as well as higher yields of plant and animal products per hectare of farmland.

Notes

1) Programma Kommunisticheskoi Partii Sovetskogo Soiuza, Gospolitizdat. 1962. p. 63.

2) V. S. Nemchinov, Ekonomiko-matematicheskie metody i modeli, Moscow. Sotsekgiz, 1962.

3) L. V. Kantorovich, Ekonomicheskii raschet nailuchshego ispol'zovaniia resursov, Moscow. USSR Academy of Sciences Press. 1959.

4) V. V. Novozhilov, "K diskussii o printsipakh planovogo tsenoobrazovaniia," in the collection Primenenie matematiki v ekonomike, Installment 1, Leningrad, Leningrad University Press, 1962, pp. 48, 49.

50

Some Functional Relations Which Arise in Analysis of a One-product Economic Model*

L. V. KANTOROVICH and L. I. GOR'KOV

In studying economic phenomena, substantial assistance may be rendered by analysis of mathematical models that enable us to study some phenomenon under simplified assumptions. We present below some simplified mathematical models that describe the process of growth of basic capital and production in a one-product dynamic model under an optimal plan.

1. Let $T(t)$ be the supply of labor at time t (a given function) and $R(t)$ be basic capital at time t. The production possibility set will be described by the technological function $U(R, T)$, which gives the quantity of net product produced by labor T using basic capital R per unit of time.

Let us look at the structure of the function $U(R, T)$. It is natural to assume it to be a homogeneous function; with this assumption we can write the function U in the form

$$U(R, T) = \int_0^n R^\alpha T^{n-\alpha} \, dp\,(\alpha), \tag{1}$$

where n is the degree of homogeneity and $p(\alpha)$ is a weight. Then all production may be considered as comprising individual "production cells" that differ in their ratios of labor to capital

---

*Doklady Akademii nauk SSSR, 1959, vol. 129, no. 4.

inputs per unit of product; in the cell corresponding to the parameter α, production is equal to $R^{\alpha}T^{n-\alpha}$. Below we will make the natural assumption that n = 1.

Thus if at time t we have labor supply $T(t)$ and basic capital $R(t)$, we produce per unit of time $U(R(t), T(t))$ units of product. This means that the function U gives optimal processes; if we assume the possibility of using linear combinations of processes, convexity of the curves $U(R, T) = \text{const}$ is required; for what follows it is enough that the function $p(\alpha)$ in (1) be nondecreasing.

Part of the produced output is used for consumption, and the remainder for accumulation. We will consider two assumptions:

a) consumption is proportional to labor supply aT; then the equation that describes the variation in capital stock may be written in the form

$$\frac{dR}{dt} = U(R(t), T(t)) - aT(t); \qquad (2)$$

b) some portion $(1 - \gamma)$ of output is consumed and the remainder is used for accumulation — addition to basic capital. In this case, we have

$$\frac{dR}{dt} = \gamma U(R(t), T(t)). \qquad (3)$$

The description of equations (2) and (3) relies heavily on the hypothesis of instantaneous transformation of capital from one form into another, to wit, into that form which is optimal under the existing ratios of capital stock and labor at a given moment.

2. We will make the model more complex, dropping the assumption of instantaneous transformation of capital from one structure into another. We will assume that production investment has a service life of α years, during which it wears out completely.

We will introduce the function $r(t, \tau)$ which gives the distribution of capital by service life at time t. Thus $r(t, \tau)d\tau$ defines the nominal capital stock [i.e., at original acquisition cost], with service life of between τ and $\tau+d\tau$ at time t. By $m(t, \tau)$ we will denote the labor force associated with this capital stock. Then the capital investment during time period $(t, t+dt)$ will be $r(t, 0)dt$. This investment comes from the portion of output used for accumulation and depreciation charges.

Output produced by capital stock $r(t, \tau)d\tau$ may be taken as equal to

$$U\left(\tfrac{1}{2}r(t, \tau),\ m(t, \tau)\right), d\tau\, dt$$

in accordance with the actual volume of capital of service life τ at time t (after deducting depreciation). To this we must add the depreciation charges, equal to $\dfrac{1}{\alpha} r(t, \tau)$, which may also be used to create new capital stock. (Incidentally, other assumptions concerning the character of depreciation and the productivity of depreciated capital may also be made.)

"Summing" over all time periods and canceling dt, we get

$$r(t,\ 0) = \gamma \int_0^\alpha U\left(\tfrac{1}{2}r(t,\ \tau),\ m(t, \tau)\right) d\tau + \int_0^\alpha \frac{r(t, \tau)}{\alpha}\, d\tau, \qquad (4)$$

where γ is the share of output going into accumulation (this equation corresponds to case b in section 1).

$R(t)$, the total actual capital stock at time t, is given by the formula

$$R(t) = \int_0^\alpha \frac{\alpha - \tau}{\alpha} r(t,\ \tau)\, d\tau. \qquad (5)$$

Total labor supply at time t must also equal

$$\int_0^\alpha m(t,\ \tau)\, d\tau = T(t).$$

Proceeding from the economic meaning of the function $r(t, \tau)$, we may assume that the condition

$$r(t,\ \tau + \Delta\tau) = r(t - \Delta\tau,\ \tau),$$

holds, or in differential form,

$$\frac{\partial r}{\partial t} + \frac{\partial r}{\partial \tau} = 0; \qquad (6)$$

similarly,

$$\frac{\partial m}{\partial t} + \frac{\partial m}{\partial \tau} = 0. \tag{7}$$

From (6) it follows that $r(t, \tau) = r(t - \tau)$ (we retain the earlier notation), and from (7) $m(t, \tau) = m(t - \tau)$. Therefore (4) and (5) may be rewritten in the following form:

$$r(t) = \gamma \int_0^{x} U(\tfrac{1}{2} r(t - \tau), \; m(t - \tau)) \, d\tau + \int_0^{\alpha} \frac{r(t - v)}{\alpha} \, d\tau. \tag{8}$$

$$R(t) = \int_0^{\alpha} \frac{\alpha - \tau}{\alpha} r(t - \tau) \, d\tau. \tag{9}$$

Differentiating (9) and putting it into (8), we get

$$\frac{dR}{dt} = \gamma \int_0^{\alpha} U(\tfrac{1}{2} r(t - \tau), \; m(t - \tau)) \, d\tau.$$

3. We will now consider a somewhat different scheme. For this we introduce the following idea. Capital has a structure u if u units of capital are required per unit of labor, so that $u = R / T$ (in other words, the structure determines the organic composition of capital).

Suppose that $\lambda(u)$ is the spectrum of distribution of labor with regard to capital, so that $\lambda(u) \, du$ is the size of the labor force associated with capital of structure from u to $u + du$; then

$$r(u) \, du = u \lambda(u) \, du \tag{10}$$

gives the capital stock of the indicated structure. The service life of capital will be assumed to be infinite — it does not wear out (and is not transformed).

We will write the equation describing the following process. Newly created product, less the share used for consumption, is used for new capital, which permits an increase in the organic composition of capital; the labor for the new capital stock comes from releasing a certain number of workers from the capital with the lowest organic structure. The capital thus released will in the future, as a rule, be unutilized.

The number of workers at all points of time is represented by a given function, i.e.,

54

$$\int\limits_{m(t)}^{M(t)} \lambda(u)\,du = T(t), \tag{11}$$

where $M(t)$ and $m(t)$ define the range of variation of the structure u at time t. Since an increase in the organic structure takes place only through newly created output (old capital is dropped),

$$r(M)\,dM = \gamma \int\limits_{m(t)}^{M(t)} U(r(u),\ \lambda(u))\,du\,dt, \tag{12}$$

where γ is the share of output going for accumulation.

The productivity of the labor force with the lowest organic composition per unit of labor is

$$\frac{U(r(m),\ \lambda(m))}{\lambda(m)}. \tag{13}$$

The effectiveness of an incremental unit of labor with the newly created capital stock is

$$\frac{\partial U}{\partial T}, \tag{14}$$

where we must take $r(M)\,dM$ for the first argument and $\lambda(M)\,dM$ as the second. Because of homogeneity, function (14) depends only on the ratio of these arguments, i.e., on M.

It is obvious that equality

$$\frac{U(r(m),\ \lambda(m))}{\lambda(m)} = \frac{\partial U}{\partial T} \tag{15}$$

must hold.

This equation, because of the homogeneity of U, has an algebraic character. If we take $U(R,T) = R^{\alpha}T^{1-\alpha}$, equation (15) takes the form

$$m = \beta M, \tag{16}$$

where $\beta = (1-\alpha)^{1/\alpha}$.

Thus the system of equations for the functions $\lambda(u)$ and $M(t)$ finally will be

$$\int\limits_{\beta M(t)}^{M(t)} \lambda(u)\, du = T(t),$$

$$M\lambda(M)\frac{dM}{dt} = \gamma \int\limits_{\gamma M(t)}^{M(t)} u^{\alpha}\lambda(u)\, du. \tag{17}$$

The system of equations of (17) loses its force if $\lambda(u)$ becomes negative.

4. All of the differential equations and integral-differential equations written above permit a stepwise numerical integration. In every case we can find different economic indicators of the model: the valuation of labor $\partial U/\partial T$, the normal effectiveness of capital investment $\partial U/\partial R$, the growth curve for basic capital $R(t)$, and the productivity of labor U/R. By comparing these variables we can study the influence of different factors (parameters of the model) and the assumptions on the variables.

In particular, we may introduce into the model the assumption of technological progress and study its influence on the economic variables. For this it is enough to replace the function $U(R, T)$, say, by $e^{\delta t}U(R, T)$ in the first case, and by the corresponding changed expressions in the two others.

Comparison of the model with data of a real economic system may permit us to reach approximate conclusions concerning some variables on the basis of other variables in such a system. The precision and justification of different methods of economic calculation may be tested through this model.

6

A One-product Dynamic Model with Instantaneous Transformation of Capital*

L. V. KANTOROVICH and I. G. GLOBENKO

1. The problem. We will consider an economic system in which one product is produced, a part being used for consumption and a part being used to increase basic capital and circulating capital. Let $T(t)$ be the labor supply available at time t (we will consider this function to be given), and let $K(t)$ be basic capital at time t (an unknown function). The possible production processes will be characterized by a function $U(K, T)$ giving net output created by labor T using basic capital K per unit of time. The function $U(K, T)$ is a positive homogeneous function of the first degree:

$$U(\lambda K, \lambda T) = \lambda U(K, T).$$

Thus if at time t we have a labor supply $T(t)$ and basic capital $K(t)$, we produce $P(t) = U(K(t), T(t))$ units of product (national income) per unit of time. It is intended that the function $U(K, T)$ be based on optimal processes. In assuming additivity and the possibility for using convex linear combinations, we are led to the requirement that the function $U(x, 1)$ $(0 \leqslant x < \infty)$ be concave below (a necessary and sufficient condition). Regarding the character of consumption, we assume

---

*Doklady Akademii nauk SSSR, 1967, vol. 174, no. 3.

that its volume $V(t)$ is given, and its size may also be determined through the parameters of the system, e.g., $V(t) = V[t, T(t), K(t), P(t)]$. Two hypotheses are usual here:

a) consumption is proportional to labor supply

$$V(t) = aT(t);$$

b) consumption depends on the volume of output, to wit, some share $(1 - \gamma)$ of total output is consumed and the remaining share γ is used for accumulation.

Development of the economy may in this case be described by a differential equation for the function $K(t)$ giving the volume of basic (and circulating) capital as a function of time

$$dK/dt = P(t) - V(t) = U(K(t), T(t)) - V[t, K(t), T(t), P(t)]. \quad (1)$$

In particular, in hypotheses a) and b) it takes the form

$$dK/dt = U[K(t), T(t)] - aT(t), \qquad (1a)$$
$$dK/dt = \gamma U[K, T]. \qquad (1b)$$

Comparison of equation (1), as well as (1a) and (1b), depends significantly on the hypothesis of instantaneous transformation of capital from one form into another, to wit, into what is optimal under the existing relations between the stock of fixed capital and labor supply at a given moment.

The development of the system is optimized at every moment of time (differential optimization), since the volume of production is determined by the state of the system and its maximum is considered in the introduction of the production function; the state of the system also determines consumption. This approach corresponds to type I one-product models discussed in [1]. Here we intend to study equation (1) and certain characteristics of the model.

The norm of effectiveness of capital investment is equal to $\partial U / \partial K$; as we know, within the assumptions of this model it represents the rate of growth of production per unit of time corresponding to a unit of incremental capital investment [2].

58

2. Integrability in explicit form. We will introduce the case of integrability of equations (1) in explicit form. We make the change of variables

$$S(t) = K(t) / T(t).$$ (2)

Equation (1) takes the form

$$S' + \frac{T'}{T} S = U(S, 1) - \frac{V}{T},$$ (3)

and equation (1b),

$$S' + \frac{T'}{T} S = \gamma U(S, 1).$$ (3b)

We will consider the following cases:

a) $T = T_0 e^{\delta t}$, $0 \leqslant t < \infty$, where δ is the rate of population growth and the function U is arbitrary;

b) $U(K, T) = cK + bT$;

c) $U(K, T) = cK^\alpha T^{1-\alpha}$ $(0 < \alpha < 1)$ (a Cobb-Douglas function);

d) $U(K, T) = K \ln T / K$ $(T \geqslant K)$.

Equation (1b) is integrated in explicit form in all cases a, b, c, and d. Equation (1a) is integrated explicitly in cases a, b, and d.

3. Formulas for the norm of effectiveness. According to the foregoing, the norm of effectiveness equals

$$n_3 = \partial U(K, T) / \partial K.$$

By introducing the variable S from (2) we get

$$n_3 = U_S'(S, 1).$$ (4)

We have

$$\frac{dP}{dt} = \frac{d}{dt} [TU(S, 1)] = T'U(S, 1) + TU_S'(S, 1) S_t'.$$

Hence, using (3), we find

$$\frac{dP}{dt} = T'U\,(S,\,1) + Tn_{9}\left[U\,(S,\,1) - \frac{T'}{T}\,S - \frac{V\,(t)}{T}\right],$$

which gives

$$n_{9} = \frac{dP/dt - T'U\,(S,\,1)}{TU\,(S,\,1) - T'S - V\,(t)} = \frac{\dfrac{1}{P}\,\dfrac{dP}{dt} - \dfrac{T'}{T}}{1 - \dfrac{T'}{T}\,\dfrac{K\,(t)}{P\,(t)} - \dfrac{V\,(t)}{P\,(t)}}. \tag{5}$$

In particular, in the case of equation (1) we get

$$n_{9} = \frac{\dfrac{1}{P}\,\dfrac{dP}{dt} - \dfrac{T'}{T}}{\gamma - \dfrac{T'}{T}\,\dfrac{K}{P}}. \tag{6}$$

From formula (6), as well as through direct computation, we get the formula for n_{9} in the case of equation (1b) and $U(K,\,T) = aU^{\alpha}T^{1-\alpha}$, to wit

$$n_{9} = a\,P\,/\,K. \tag{6'}$$

4. Consideration of technological progress. Technological progress in equation (1) is considered in a very simple way. If for production we use the formula

$$P\,(t) = e^{\rho t}U\,(K,\,T),$$

we get

$$K' = e^{\rho t}U\,(K,\,T) - V\,(t), \tag{7}$$

where:

a) $V\,(t) = aT\,(t)$;

b) $V\,(t) = (1 - \gamma)\,e^{\rho t}U\,(K,\,T),\, 0 < \gamma < 1$.

We now calculate the norm of effectiveness. We have

$$\frac{dP}{dt} = \frac{d}{dt}\,[e^{\rho t}U\,(K,\,T)] = \rho e^{\rho t}U + e^{\rho t}\frac{d}{dt}\,[T\,(t)\,U\,(S;\,1)] =$$
$$= \rho e^{\rho t}TU\,(S,\,1) + e^{\rho t}T'U\,(S,\,1) + e^{\rho t}TU'_{S}\,(S,\,1)\,S'_{t};$$

A One-product Dynamic Model

but from (7)

$$S' = e^{\rho t}U\,(S,\,1) - \frac{T'}{T}\,S - \frac{V}{T}.$$

In this case $n_3 = e^{\rho t}U_S'(S,1)$, so that we get

$$\frac{dP}{dt} = \rho e^{\rho t}TU\,(S,\,1) + e^{\rho t}T'U\,(S,\,1) + Tn_3\left[e^{\rho t}U\,(S,1) - \frac{T'}{T}\,S - \frac{V}{T}\right],$$

$$n_3 = \frac{\dfrac{dP}{dt} - \rho P - e^{\rho t}T'U\,(S,\,1)}{T\left(e^{\rho t}U\,(S,\,1) - \dfrac{T'}{T}\,S - \dfrac{V}{T}\right)} = \frac{\dfrac{1}{P}\dfrac{dP}{dt} - \left(\rho + \dfrac{T'}{T}\right)}{1 - \dfrac{T'}{T}\dfrac{K}{P} - \dfrac{V}{P}}.$$

5. Calculation of depreciation and obsolescence. The equation for the depreciated (actual), but not nominal, capital stock, denoted $\overline{K}$, will be

$$\frac{d\overline{K}}{dt} = U\,(\overline{K},\,T) - \delta\overline{K}\,(t) - V\,(t),$$

where δ is the share of capital lost because of physical wear, obsolescence, and failure of capital to correspond to the necessary structure of production. For this case we apply the general formula (the term $\delta\overline{K}(t)$ also corresponds to the corresponding increase of consumption). Therefore

$$n_3 = \frac{\dfrac{1}{P}\dfrac{dP}{dt} - \dfrac{T'}{T}}{1 - \dfrac{V}{P} - \dfrac{\overline{K}}{P}\dfrac{T'}{T} - \delta\dfrac{\overline{K}}{P}}.$$

To take account of the length of the construction period the equation would have to be written differently. Insofar as it is desired to increase the existing capital stock, we must consider depreciation and obsolescence. With a period of immobilization of resources in construction equal to v years, if we assume a smooth character for the function in the equation, the equation which determines the variation in $\overline{K}$ (the depreciation) may be written approximately as:

$$(1 + \beta)^\nu \frac{d\overline{K}}{dt} = U\left(\overline{K}(t),\, T(t)\right) - \delta\overline{K}(t) - V(t);$$

$$\beta = \frac{1}{P}\frac{dP}{dt} \quad \text{or} \quad \beta = \frac{1}{P}\frac{dP}{dt} + \delta.$$

Therefore,

$$n_\vartheta = \frac{\left(\dfrac{1}{P}\dfrac{dP}{dt} - \dfrac{T'}{T}\right)(1+\beta)^\nu}{1 - \delta\dfrac{\overline{K}}{P} - \dfrac{V}{P} - \dfrac{T'}{T}\dfrac{\overline{K}}{P}(1+\beta)^\nu}.$$

It would be more accurate to proceed from analysis of an equation with a lag. The influence of economic wear may be studied by the type III model in [1].

6. <u>Asymptotic behavior</u>. We will look at equation (1b); assuming that $\lim\limits_{t\to\infty} \dfrac{T'}{T} = \lambda$, we may state the following theorems:

<u>Theorem 1.</u> Suppose that c is a root of the equation

$$\lambda x = \gamma U(x, 1), \quad 0 < x < \infty;$$

K is a solution of equation (1b); then we have

$$\lim_{t\to\infty} \frac{K}{T} = c, \quad \bar{n}_\vartheta = \lim_{t\to\infty} n_\vartheta = U'_x(c, 1).$$

<u>Theorem 2.</u> Suppose that $\gamma U(x, 1) > \lambda x$ for $0 < x < \infty$; K is a solution of equation (1); then we have

$$\lim_{t\to\infty} \frac{\ln K}{t} = a\gamma, \quad \lim_{t\to\infty} n_\vartheta = \lim_{t\to\infty} \frac{U(K, T)}{K} = a,$$

where

$$a = \lim_{x\to\infty} U'_x(x, 1).$$

Taking into account technological progress, we get

<u>Theorem 3.</u> Suppose that in the neighborhood of the origin

A One-product Dynamic Model

$$t^{\alpha-1}U(1, t) = [c_0 + O(t)]$$

and suppose that there exist limits for $n_\mathfrak{z}$ and for $xU_x'(x, 1) / U(x, 1)$ as $x \to \infty$. Then if K is a solution of equation (7), the following asymptotic formulas hold:

$$\lim_{t \to \infty} \frac{K}{e^{\frac{\rho}{1-\alpha}t} T} = \left[\frac{(1-\alpha) c_0 \gamma}{\rho + (1-\alpha) \lambda} \right]^{1/(1-\alpha)},$$

$$\lim_{t \to \infty} n_\mathfrak{z} = \alpha \frac{\rho + (1-\alpha) \lambda}{(1-\alpha) \gamma}.$$

Theorems like 1, 2, and 3 may be formulated and proved in the case of model (1a) as well.

7. A numerical example. To illustrate the foregoing formulas we will present a numerical example based on certain assumed date. We will assume that in the existing state of the economy the rate of growth of national income $\frac{1}{P}\frac{dP}{dt} = 0.08$; $T'/T = 0.02$; $\frac{V}{P} = 0.7$. From (6) $n_\mathfrak{z} = 0.23$, i.e., 23%. Formula (6') would give the same value if $\alpha = 0.46$ (according to data for the U.S. economy, $\alpha = 0.33$). Taking into account technological progress with $\rho = 0.02$, we have $n_\mathfrak{z} = 15.4\%$. If we take into account depreciation with $\delta = 0.03$, we get $n_\mathfrak{z} = 30\%$ if we assume that $\overline{K} = K$, and $n_\mathfrak{z} = 27\%$ if $\overline{K} = 0.75 K$. If we make both assumptions for the same ρ and δ, we get $n_\mathfrak{z} = 20\%$. Finally, if we assume a construction period of $\nu = 2$ years, take $\overline{K} = 0.75 K$ and $\beta = \frac{1}{P}\frac{dP}{dt}$, we have $n_\mathfrak{z} = 0.32$.

References

1. L. V. Kantorovich and L. I. Gor'kov, DAN, 1959, vol. 129, no. 24.
2. L. V. Kantorovich and V. L. Makarov, Primenenie matematiki v ekonomicheskikh issledovaniiakh, 1965, no. 3, pp. 70-72.

7

A Dynamic Model of the Economy*

L. V. KANTOROVICH and I. G. GLOBENKO

1. In [1] we considered a one-product model of the economy and used it to calculate the norm of effectiveness. A more specific analysis may investigate the behavior of a number of industries and products. In the present paper, retaining the macroeconomic character of the model, we introduce two products and two departments, following the scheme of Karl Marx. It is assumed that the output of the first department serves as the source of capital for both departments, and the output of the second department is used for consumption. The conditions of production, which may be expressed by a production function, are not assumed to be identical for both departments. We will denote by K_1 and K_2 the capital stock of the first and second departments, and by P_1 and P_2 the net output of the first and second departments. We assume that production functions U_1 and U_2 are given. Then

$$P_1 = U_1[K_1,\ T_1],\quad P_2 = U_2[K_2,\ T_2],$$

where T_1 and T_2 are the labor resources used in the first and second departments, respectively. Here $T_1 + T_2 = T$, and the labor supplies of the two industries are determined, for ex-

---

*Doklady Akademii nauk SSSR, 1967, vol. 176, no. 5.

ample, by the demographic growth rate; the functions $U_i(x,y)$ $(i = 1,\ 2)$ are positive, homogeneous, first degree, and twice differentiable, with

$$U_i{}'(x,\ 1) > 0, \quad U_i{}''(x,\ 1) < 0 \quad (i = 1, 2), \quad 0 < x < +\infty,$$
$$U_i(0,\ 1) = U_i(1,\ 0) = 0 \quad (i = 1,\ 2).$$

Consumption $V(t)$ is given in advance or represents a known function of the parameters of the system, and accumulation (the output of the first department P_1) is divided between the first and second departments.

We make the following assumptions: capital in each department may be transformed instantaneously, but it may not be transferred to the other department, so that the capital stock in each department cannot be reduced. Labor may be transferred.

Thus we have the relations:

$$T_1 + T_2 = T; \tag{1}$$
$$U_2(K_2,\ T_2) = V(t); \tag{2}$$
$$K_1{}' + K_2{}' = U_1(K_1,\ T_1). \tag{3}$$

Further, the nontransferability of capital gives the requirement that $K_1(t)$ and $K_2(t)$ be nondecreasing functions.

Finally, differential optimization requires that the condition of equally efficient distribution of labor and capital hold

$$\frac{\partial U_1}{\partial K_1}\frac{\partial U_2}{\partial T_2} = \frac{\partial U_1}{\partial T_1}\frac{\partial U_2}{\partial K_2}, \quad \text{if} \quad K_1{}' > 0, \quad K_2{}' > 0. \tag{4}$$

The size of the norm of effectiveness determines the growth of output per unit of incremental capital investment per unit of time and is equal to

$$n_3 = \partial P_1 / \partial K_1.$$

The following cases of integrability hold:
1) $U_1(K_1,\ T_1) = K_1{}^{\alpha}T_1^{1-\alpha}$, $U_2 = K_2{}^{\beta}T_2^{1-\beta}$: the Cobb-Douglas case.

65

2) $U_1(K_1, T_1) = a_1 K_1 + b_1 T_1$, $U_2(K_2, T_2) = a_2 K_2 + b_2 T_2$.

We note that if $T_2(t)$ is given, $K_2(t)$ is also determined, and we find ourselves in the situation of the one-product model of [1]. The case of identical functions U_1 and U_2 also boils down to a one-product model.

2. By Euler's Theorem we have

$$\frac{\partial U_1}{\partial K_1} K_1 + \frac{\partial U_1}{\partial T_1} T_1 = U_1, \quad \frac{\partial U_2}{\partial K_2} K_2 + \frac{\partial U_2}{\partial T_2} T_2 = U_2.$$

On the basis of (2) and (3) we find

$$\frac{\partial U_1}{\partial K_1} K_1' + \frac{\partial U_1}{\partial T_1} T_1' = U_{1t}', \quad \frac{\partial U_2}{\partial K_2} K_2' + \frac{\partial U_2}{\partial T_2} T_2' = V'(t), \quad K_1' + K_2' = U_1.$$

Hence, still using (4), we get the formula

$$n_3 = \frac{V(U_1 T_1' - U_{1t}' T_1) + U_1(V T_2' - V' T_2)}{V(k_1 T' - U_1 T_1) + V'(K_2 T_1 - K_1 T_2)},$$

which in the Cobb-Douglas case takes the form

$$n_3 = a U_1 / K_1.$$

Taking into account technological progress, the expression for n_3 takes the form

$$n_3 = \frac{P_2(P_1 T_1' - P_{1t}' T_1 + \rho_1 P_1 T_1) + P_1(P_2 T_2' - P_{2t}' T_2 + \rho_2 P_2 T_2)}{P_2(K_1 T' - P_1 T_1) + (P_{2t}' - \rho_2 P_2)(K_2 T_1 - K_1 T_2)},$$

where $P_1 = e^{\rho_1 t} U_1(K_1, T_1)$, $P_2 = e^{\rho_2 t} U_2(K_2, T_2)$, $\rho_1, \rho_2 \geqslant 0$.

3. We will assume that there exist the limits

$$\lim_{t \to \infty} T_1 / T = a \neq 0; \tag{5}$$

$$\lim_{t \to \infty} T' / T = \lambda; \tag{6}$$

$$\lim_{t \to \infty} V(t) / T(t) = b. \tag{7}$$

A Dynamic Model of the Economy

If (5)-(7) hold, it can be shown that $b < +\infty$. Furthermore, the following cases are possible:

1) The straight line $y = \lambda x$ and the curved line $y = U_1(x - c_2, a)$ intersect at two points $c_0 < c_1$ (c_2 is a root of the equation $b = U_2(x, 1 - a)$, $c_2 \leqslant x < +\infty$).

The following formulas hold:

$$\lim_{t \to \infty} K_2/T = c_2; \tag{8}$$

$$\lim_{t \to \infty} \frac{K_1}{T} = \begin{cases} \text{a)} & c_0 - c_2, \\ \text{b)} & c_1 - c_2. \end{cases} \tag{9}$$

Here, if a constant $c > c_0$ exists such that we can find an arbitrarily distant value of t for which $s(t) \geqslant c$, we have case a), and otherwise, case b).

2) The straight line $y = \lambda x$ and the curved line $y = U_1(x - c_2, a)$ have one point in common. In this case formulas (8) and (9) hold and cases a) and b) coincide.

3) The straight line $y = \lambda x$ and the curved line $y = U_1(x - c_2, a)$ do not have points in common. In this case, starting from some moment of time, K_1 and K_2 become negative and the model loses its economic meaning.

If (5)-(7) hold, we have the asymptotic formulas

$$\lim_{t \to \infty} n_\partial = U_1'\left(\frac{c_0 - c_2}{a}, 1\right) \quad \text{for case a);}$$

$$\lim_{t \to \infty} n_\partial = U_1'\left(\frac{c_1 - c_2}{a}, 1\right), \quad \text{for case b);}$$

The function U_1' means $U_{1x}'(x, 1)$.

Reference

L. V. Kantorovich and I. G. Globenko, DAN, vol. 174, 1967, no. 3.

8

On the Calculation of the Norm of Effectiveness on the Basis of a One-product Model of the Development of the Economy*

L. V. KANTOROVICH and Al'b. L. VAINSHTEIN

1. The Appearance and Significance of Mathematical Macroeconomic Models for Analysis of the Long-range Development of the National Economy

Study of the concrete dynamics of the national economy as a whole began in bourgeois economic science, properly speaking, with the study of the crises that arose in the middle of the nineteenth century and that required both explanation and prediction. But the crisis was studied as a separate moment of time, as a separate point in the dynamics of the national economy that went beyond the range of the usual or feasible.

Karl Marx's ingenious models of reproduction in the second volume of Capital were developed on the basis of arithmetical examples and did not represent in essence a mathematical model, although Marx, as we know, often stressed the need for using mathematical methods in economic research. In particular, Marx wanted to mathematically derive the main laws of crises, but then he rejected this objective (see Marx's 1873 letter to Friedrich Engels, Sobr. soch., Vol. 33).

Only at the start of the twentieth century, at the persistent

*Ekonomika i matematicheskie metody, 1967, no. 5.

urging of W. Sombart, and then G. Cassel, was the study of
crises replaced by the study of the business cycle, i.e., the economic cycle as a whole. However, here the object of study was
not the total dynamics but only the business cycle, its phases,
and the position in it of individual national economic indicators
at any given moment. Trends in the development of the national
economy as a whole were not then of interest to the investigator.
On the contrary, a distinct evolutionary curve hindered him; he
strived to exclude it from the overall economic series in order
to get at the cyclical fluctuations in "pure" form and for individual partial indicators.

Such was the methodological approach to the study of dynamics of the national economy of the Anglo-American statistical
school (Persons, Yule, and others); on this basis, in the first
quarter of the twentieth century numerous "economic barometers" that were supposed to forecast the "economic weather"
sprang up in profusion.

However, because of the inconsistency of these barometers,
and under the influence of the unprecedented crisis that seized
the entire capitalist world during the 1930s, it became necessary for investigators to study the overall dynamics of the
national economy, its regularities, and the interdependence of
individual variables: to study the theory of harmonic and stable
economic growth. Such an approach might be called a macroeconomic approach, and John Keynes was its primary exponent
in the foreign economic literature. At the center of his analysis
he put synthetic national indicators such as national income,
total consumption, total savings, total supply, total demand, and
so on, as well as the behavior of certain marginal variables —
parameters of his macroeconomic model: the marginal propensity to consume, the marginal propensity to save, the marginal
efficiency of capital, the multiplier, the accelerator, and so
on [1]. Proceeding from certain assumptions concerning the
necessary relation between individual variables (total savings,
consumption, capital investment, and so on) and the assumed
behavior of certain parameters in their interrelationship,
Keynes established the conditions under which his model of

the national economy could develop in a stable manner.

The work of Keynes received vast attention throughout the capitalist world and in the theoretical literature. The overwhelming majority of foreign authors who concern themselves with the theory of growth and the construction of dynamic economic models (Harrod and Domar, Philips, Samuelson, Hicks, and a number of others) follow in the wake of Keynes: they change the suggested equations somewhat; they introduce lags in the variables (for example, output-income, income-expenditure, expenditure-output, where the second element is lagged behind the first); they introduce lags in the multiplier or the accelerator, or both. Many models of a modified Keynesian type may be found in R. Allen's book [2], especially Chapters 2 and 3, although under the conditions of the capitalist economy they may have only a theoretical significance.

However, priority in the construction of a macroeconomic dynamic model and the formulation of the basic conditions for stable and constant growth of the economy belongs not to Keynes but to the Soviet economist and engineer, the famous specialist of the USSR State Planning Commission, G. A. Fel'dman, who as early as 1928 published two articles in which he presented his theory of the rates of growth of national income [3]. This major work, as well as certain other interesting ideas of Russian economists, remained unknown for a long time, and only after the Second World War did it become known abroad and provoke vast interest. In 1957 the famous American econometrician E. Domar, in his book on the theory of economic growth, devoted one of his essays completely to a very detailed analysis of the articles of Fel'dman, terming them the remarkable beginning of the creation of the mathematical theory of growth [4, Essay 9]. In 1964 Fel'dman's work was published in full in the USA [5], and one year earlier, in Poland. (1)

In his work Fel'dman divides all production into two parts, but not according to Marx's scheme: one part serves for the production of consumer goods, as well as means of production necessary to maintain the volume of production of consumer goods at each given level; the other part serves to increase and replace production capital. Then he establishes the mathematical relationships between different components and elements of these outputs and the rates of their growth for three cases of economic development: (1) total consumption is constant; (2) consumption increases at a constant relative rate; and (3) consumption increases at an increasing rate. The author determines the effect of the change in any factor on all the other elements of the economy in order to control it in the desired manner. For example, he determines the influence of a simultaneous change of equipment in industry, which in the capitalist economy inevitably provokes sharp cyclical fluctuations that can be avoided in a planned economy. Or he studies the parameters and how they change in order to double the rate of growth of consumption; in a word, Fel'dman's model is a model of a controlled economy that is intended precisely for the planned national economy.

Fel'dman's model and his theory of growth formed the basis for the long-range (15-20 years) plan for the development of the national economy of the USSR produced in the late 1920s by Gosplan under the direction of N. Kovaleskii, but then this work was interrupted.

Construction of macroeconomic models which could be applied in the controlled socialist economy was resumed only in the second half of the 1950s, after the discovery of linear programming in the USSR and the start of the extensive application of mathematical methods in economic research. In the work of L. V. Kantorovich and L. I. Gor'kov in 1959 [7], several such models were constructed: models that assume instantaneous transformation of capital, that assume unconvertibility of capital and take into account depreciation; models which gradually write off capital in connection with its economic aging. These models lead to differential equations, to rather original integral-differential equations, and to other functional relationships. This work was the subject of an address by L. I. Gor'kov at a conference in 1960 [8] and it elicited interest; in particular, A. N. Kolmogorov participated in the discussion, and he recognized the usefulness of such simplified models. However, work

71

in this direction was actually interrupted and was only recently revived by L. V. Kantorovich in cooperation with I. G. Globenko. Some results from it were published that relate to a simplified one-product model with instantaneous transformation of capital [10].

It should be noted that in the foreign literature a great many works have recently been devoted to optimal growth models, but they have been constructed on a somewhat different level (for example, Koopmans, Gale, Inagaki [11-13]). In them the problem of optimization of the distribution of capital between the present and the future is raised, and it is done in an extremely conditional and indeterminate formulation. In our work this question is not considered, and the basis for the distribution of the net product between consumption and accumulation is assumed to be given.

2. Construction of a General Model of Development

The subject of our paper is also a one-product model. At first glance a one-product model is an extreme stage of abstraction in relation to the real national economy. Only one product is considered: it is eaten, it is used for clothing, it is used to construct factories and to make machines. In fact this abstraction is not so great in actuality, and under an economic approach all products may be commensurated — in money or in labor (in accordance with the adopted conception of price formation). We recall that Marx's two-product model of expanded reproduction is the same kind of abstraction, but it nevertheless provides extensive possibilities for meaningful analysis of the economy. Valuing each product in the same units, we can consider any model of the national economy as a one-product model, and from the mathematical viewpoint the method by which all products may be commensurated and transformed into one — through money, labor, energy, or any other method — is not so important.

Thus an economic system is considered in which one product

72

is produced. Part of it is used for consumption, and part to increase basic and circulating capital. In this model these two kinds of capital are not distinguished.

The basic role in the model is played by the production function $P(t)$, which describes the amount of net product or national income that can be produced per unit of time with a capital stock of size $K(t)$ and labor supply of $T(t)$ (2), available at time t (labor is all reduced to a single type) or

$$P(t) = U[K(t),\ T(t)]. \qquad (1)$$

Here the function $T(t)$ is assumed to be given, determined on the basis of the dynamics of the labor force employed in the national economy, which, in its turn, is permitted to changes with population. $K(t)$ represents all the capital in the national economy — basic and circulating capital available at time t. This function is an unknown, and the initial $K(0)$ is given.

Two quite natural assumptions underlie the construction of the model.

The first assumption. The function $U(K, T)$ already includes an optimal solution, i.e., it is known how it is necessary to arrange in the best possible way the existing volumes of capital and labor in order to produce the greatest quantity of product per unit of time.

The second assumption. The function U is positive and homogeneous, i.e., if we have λ times as much labor and capital, we will be able to produce λ times as much output — $U(\lambda K,\ \lambda T) = = \lambda U(K, T)$.

From the foregoing assumptions that the function $U(K,\ T)$ is based on optimal methods and that the methods may be combined linearly, it follows that together with every two points in the set $U(K, T) \geqslant C$ there is an intermediate point, and consequently, the function $U(x, 1)$ turns out to be concave below.

The total (3) national consumption $V(t)$ may be considered as given at a given moment of time, or we can assume that this function is determined by parameters of the system; i.e.:

$$V(t) = V[t, T(t),\ K(t),\ P(t)]. \qquad (2)$$

Thus knowing the resources that the economy has at time t and what output is produced, we determine total national consumption.

Then the growth of the economy (the rate of increase of capital) may be described by a very simple differential equation based on (1) and (2):

$$\frac{dK}{dt} = P(t) - V(t) = U[K(t), T(t)] - V[t, T(t), K(t), P(t)] \quad (3)$$

For the form of the consumption function we may make, in particular, one of the following simplified assumptions:

a) consumption is proportional to labor supply $V(t) = \alpha T(t)$;

b) a certain share of total product $(1 - \gamma)$ is used for consumption, and the remaining part γ goes for accumulation.

Making such assumptions about consumption, equation (3) gives

$$\frac{dK}{dt} = U[K(t), T(t)] - \alpha T(t), \quad (3a)$$

$$\frac{dK}{dt} = \gamma U[K(t), T(t)]. \quad (3b)$$

Equations (3), (3a), and (3b) depend heavily on the assumption of instantaneous transformation of capital, i.e., we are assuming that no matter what the physical form of capital, it may easily be transformed into another form. For example, if we do not want old tractors, they can be sold and for the same amount we can buy new ones, i.e., the producer always has the desired structure of capital, which is the optimal structure, for optimization is included in the very construction of the production function. This form of the production function and its application rely not only on the assumption of instantaneous transformation of capital but also on the assumption of optimal utilization of existing resources at every moment of time.

In our system consumption is determined by the parameters, so that in fact we have here a determinate equation such that there are no control parameters or optimization. Never-

74

theless, this optimization takes place implicitly through the continuous optimal transformation of capital and the use of the production function that corresponds to the optimal choice of production processes. Therefore we may say that this is differential optimization in a certain sense, i.e., at each moment of time we choose that policy which gives the greatest increase in capital at the given moment, considering the consumption requirements that must be satisfied.

3. The Norm of Effectiveness of Capital Investment and Its Mathematical Formulation

Our model relates to the first of the three kinds of models that have figured in the work of Kantorovich and Gor'kov [7]. In studying the model we will be interested in one parameter of the system — the norm of effectiveness of capital investment. The norm of effectiveness is extremely important for economic research, since to calculate the effectiveness of specific capital investments and to calculate prices as prices of production and by the scheme of optimal planning, we must know this parameter.

However, there is no objective approach to the determination of the norm of interest to us. The method for selecting capital investments applied by V. V. Novozhilov and L. V. Kantorovich has a somewhat theoretical character, and it is extremely difficult to apply it to specific materials. Thus the need to calculate the norm remains quite acute. In such an important question as the determination of new prices (as of July 1967), a norm of effectiveness of 15% was adopted very tentatively without theoretical justification. Therefore we must determine an objective method to determine this basic national economic indicator.

It seems to us that the method of macroeconomic models, especially that of the preceding section, permits some quantitative estimation of the norm that interests us and a qualitative investigation of this indicator and its dependence on different parameters of the economic system.

To investigate it, it is useful in equation (3) to replace the variables and introduce a new unknown, to wit, the volume of capital per unit of available labor (the capital-labor ratio)

$$S(t) = \frac{K(t)}{T(t)}. \tag{4}$$

Then equation (3) after this substitution takes the form:

$$S' + \frac{T'}{T} S = U(S, 1) - \frac{V}{T} \tag{5}$$

and the special case (3b)

$$S' + \frac{T'}{T} S = \gamma U(S, 1).$$

We note the following cases in which the equation is integrated in final form in elementary functions or in quadratures, i.e., in analytic form:

1) when $T = T_0 e^{\delta t} (0 \leqslant t < \infty)$, i.e., it is assumed that there is an increase in labor or national population at a constant relative rate (an exponential function);

2) $U(K, T) = cK + bT$, i.e., the linear case. It is of interest since any function may be approximated over some particular interval;

3) $U(K, T) = dK^{\alpha}T^{1-\alpha}$, $0 < \alpha < 1$. This is a special case of the Cobb-Douglas function. (4) Through American statistics it provides a good description of the production of enterprises with different volumes of capital and labor;

4) $U(K, T) = K \ln (T / K)$, $T \geqslant K$.

Continuous models give results in analytical form which simplifies qualitative analysis. But we may proceed by another route — construction of discrete dynamic models. This approach has certain advantages for the selection of parameters and provides an opportunity for complex structures. (5) The two methods of analysis must, obviously, not contradict each other, and it is wise to combine them.

We now turn to the basic parameter of interest to us — the norm of effectiveness of capital investment. The norm of effectiveness for our national economy is that growth in net output is generated per unit of time by a properly utilized incremental unit of capital investment. This definition has a very real meaning only in an optimally controlled economy.

The norm of effectiveness under these assumptions is expressed by the following simple formula (6):

$$\eta_{\ni} = \frac{\partial U(K, T)}{\partial K}. \tag{6}$$

We now express $\eta_{\ni}$ through other variables and parameters, but first we make the change of variables $S = K / T$ and $K = ST$. Then (6) is written as

$$\eta_{\ni} = U_S'(S, 1). \tag{6a}$$

We now write the equation for growth of national income from (1):

$$\frac{dP}{dt} = \frac{d}{dt}[TU(S,1)] = T'U(S,1) + TU_S'(S,1)S_t'.$$

We replace $U_S'(S, 1)$ in this equation by $\eta_{\ni}$ and solve the resulting equation for the norm of effectiveness $\eta_{\ni}$. Then we get (using (4) and (5))

$$\eta_{\ni} = \frac{\dfrac{dP}{dt} - T'U(S,1)}{TS_t'} = \frac{\dfrac{dP}{dt} - T'U(S,1)}{TU(S,1) - T'S - V} =$$

$$= \frac{\dfrac{1}{P}\dfrac{dP}{dt} - \dfrac{T'}{T}}{1 - \dfrac{V}{P} - \dfrac{T'}{T}\dfrac{K}{P}} \tag{7}$$

All the variables in formula (7) have a very clear economic meaning and are known to every economist:

in the numerator we have the rate of growth of national income and the rate of growth of labor (T' / T);

in the denominator we have the ratio of capital to national income, or the capital-output ratio (K / P), and the share of consumption in the national income (V / P).

In particular, in the case of the Cobb-Douglas function we get the extremely simple formula

$$\eta_\vartheta = a \frac{P}{K}. \tag{7a}$$

From formula (7) we may derive other results, such as the relationship between the rate of growth of national income and various factors:

$$\frac{1}{P} \frac{dP}{dt} = \frac{T'}{T} + \eta_\vartheta \left(1 - \frac{T'}{T} \frac{K}{P} - \frac{V}{P} \right).$$

The structure of formula (7) warrants attention: the production function does not enter explicitly the expression for the norm of effectiveness, and η_ϑ is related directly to global indicators of the national economy.

4. Modification of the Model

1. Consideration of technological progress. Technological progress may be interpreted in the following way: after a few years, with a capital stock of the same size (of a given value) and with the same labor input, we can produce more output. The formula for national income (1) will then have the following form:

$$P(t) = e^{\rho t} U(K, T).$$

The equation given earlier (3) will change correspondingly

$$K' = e^{\rho t} U(K, T) - V(t).$$

Without repeating all the details we can write the formula for the norm of effectiveness taking into account technological progress:

$$\eta_\mathfrak{z} = \left(\frac{1}{P} \frac{dP}{dt} - \left(\rho + \frac{T'}{T} \right) \right) \Big/ \left(1 - \frac{V}{P} - \frac{T'}{T} \frac{K}{P} \right). \tag{8}$$

2. Consideration of physical wear and obsolescence. Taking into account the physical wear and obsolescence of capital, we can bring the model somewhat closer to reality. The capital stock may be altered only to a slight extent, but to some degree it does become obsolete or worn out. We must also assume that the capital that has not actually become obsolete or physically worn becomes unnecessary and ceases to correspond to the required structure of output (for example, blocks for women's hats and lasts for outmoded shoe styles). For all these reasons some part of the capital stock is discarded annually.

We will denote by δ the share of capital discarded annually for all reasons. To take depreciation into account, we should use a differential equation, not for the capital stock in terms of nominal value (the original acquisition cost) but for the depreciated value of the capital stock $\overline{K}$. Then equation (3) may be written in the form (7):

$$\frac{d\overline{K}}{dt} = U[\overline{K}(t), T(t)] - \delta \overline{K}(t) - V(t)$$

and the formula for the norm of effectiveness

$$\eta_\mathfrak{z} = \left(\frac{1}{P} \frac{dP}{dt} - \frac{T'}{T} \right) \Big/ \left(1 - \frac{V}{P} - \frac{T'}{T} \frac{\overline{K}}{P} - \delta \frac{\overline{K}}{P} \right). \tag{9}$$

3. Consideration of the period for creating capital lags. Accumulation does not instantaneously become capital; it takes several years to build and assimilate plants. The construction periods may differ, and capital to a certain extent becomes obsolete during the time of its construction. This may be taken into account in two ways. First, we can consider capital in different periods and derive differential lagged equations — we will not consider this approach here.

The second procedure is to assume that the capital introduced today, for example, was created from accumulation two years ago. Suppose that national income is 1 billion rubles and that two years ago it was 800 million rubles. If we took a certain proportion for accumulation, the fact that this share was taken out of the accumulation of two years earlier will reduce it by 25% (the growth of national income over the two years is 25%). During the two years, in addition, there was some obsolescence. For this case we may write the following heuristic, somewhat tentative equation if the functions entering it are smooth:

$$(1 + \beta)^{\mu} \frac{d\overline{K}}{dt} = U(\overline{K}, T) - \delta \overline{K} - V,$$

where μ is the average period for the realization of accumulations (the weighted average period of immobilization of capital during the construction process added to the "discounted" period of assimilation [8]). Since the capital coming on stream in a given year is determined by the accumulations made μ years earlier, this accumulation is less than the accumulation of a given year by an amount $(1 + \beta)^{\mu}$, whatever the multiplier in the preceding formula.

For β we may use one of two formulas:

$$\beta = \frac{1}{P} \frac{dP}{dt},$$

i.e., the change in accumulation is taken in conformity with the rate of growth of the national economy; or

$$\beta = \frac{1}{P}\frac{dP}{dt} + \delta.$$

Here, besides the lag we also consider the obsolescence of capital during the course of construction.

When taking into account the construction period for capital, the effectiveness formula takes the form

$$\eta_\mathfrak{s} = \left[\left(\frac{1}{P}\frac{dP}{dt} - \frac{T'}{T}\right)(1+\beta)^\mu\right]$$

$$\Big/\left[1 - \frac{V}{P} - \delta\frac{\overline{K}}{P} - \frac{T'}{T}\frac{\overline{K}}{P}(1+\beta)^\mu\right].$$

(10)

5. Calculation of the Norm of Effectiveness

The assumptions underlying this model — instantaneous transformation of capital and optimal decision-making not only from the very start but during the entire interval of time studied — are actually realized under our specific conditions only to a certain extent, so that the values calculated according to the formulas that have been presented do not yet completely reflect the real norm of effectiveness (more about this later). However, their numerical expression, if we put into the formulas statistical data that represent the state of the national economy of the USSR today, may have some orientational value. In the future, assumptions may be considered that take better account of reality, and the result will be more reliable.

The initial absolute figures necessary to calculate the norm of effectiveness numerically will be based on data of the USSR Central Statistical Administration, with the value indicators taken in comparable prices except in cases specifically stipulated. The relative values are calculated by us through analysis of trends corresponding to time series of five or ten years.

The following observation must be made in connection with the figures given below. The formulas given in the preceding section relate to net output or national income of the USSR,

Table 1

Item No.	Measure	Symbol	Numerical Value	Observations
1	National income for 1965	P	203.4 billion rubles	Produced income; Narodnoe khoziaistvo v 1965, p. 589
2	Share of consumption in national income	V/P	0.74	Average value taken from A. L. Vainshtein [15]
3	Value of basic and circulating capital in the national economy as a whole	K	620 billion rubles	As of January 1, 1965, at total replacement cost
4	The same, allowing for wear	$\overline{K}$	498 billion rubles	Depreciation applied only to basic capital, 26%, according to data of the USSR Central Statistical Administration. For circulating capital the CSA does not calculate wear. In general, $\overline{K} = 0.80K$
5	Value of basic and circulating capital in area of material production	K_0	423 billion rubles	2.6% of circulating capital is imputed to nonmaterial production sphere, according to data for January 1, 1966
6	The same, allowing for wear	$\overline{K}_0$	351 billion rubles	Wear calculated only on basic capital

82

Item No.	Measure	Symbol	Numerical Value	Observations
7	Capital-output ratio (acquisition cost basis)	K/P	$3.05 \approx 3.0$	Line 3 divided by line 1
8	Capital-output ratio (actual basis)	$\bar{K}/P$	2.45	Line 4 divided by line 1
9	Capital-output ratio in production sphere (acquisition cost basis)	K_0/P	$2.08 \approx 2.1$	Line 5 divided by line 1
10	Capital-output ratio in material production sphere (actual basis)	$\bar{K}_0/P$	1.72	Line 6 divided by line 1
11	Rate of growth of national income	$\dfrac{1}{P}\dfrac{dP}{dt}$	$0.0653 \approx 6.5\%$	Rough analysis of 1950-64 growth in work by A. L. Vainshtein [15]
12	Average rate of growth of labor supply in national economy	T'/T	0.024 or 2.4%	Approximation for 1966 based on smoothed linear trend of employment in all sectors, including collective farms (1959-66)
13	Average rate of growth of labor supply in material production sphere	T_0'/T_0	0.018 or 1.8%	As 12 applied only to material production sphere

Table 2

Formula	$\eta_\text{э}$	$\eta_\text{эо}$	Observations
(7) Basic formula $$\dfrac{\dfrac{1}{P}\dfrac{dP}{dt} - \dfrac{T'}{T}}{\dfrac{V}{P} - \dfrac{T'}{T}\dfrac{K}{P}}$$	0.22	0.21	By Cobb-Douglas formula (7a), assuming calculated norm of effectiveness of (0.22) we get $\alpha = 0.67$ — which is rather high
(7a) Cobb-Douglas, $\alpha P/K$ for the USSR	0.13	0.183	$\alpha = 0.382$. Calculated through work by B. N. Mikhalevskii and Iu. P. Solov'ev [17], in which the authors give a three-factor production function for 1951-63
(8) Model takes into account technological progress $$\dfrac{\dfrac{1}{P}\dfrac{dP}{dt} - \left(\rho + \dfrac{T'}{T}\right)}{\dfrac{V}{P} - \dfrac{T'}{T}\dfrac{K}{P}}$$	0.146	0.159	$\rho = 0.014$ and $\rho_0 = 0.013$; they are calculated according to formula (11), and P_2/P_1, K_2/K_1 and T_2/T_1 are calculated according to data for 1966 and 1965 and are equal, respectively, to 1.065, 1.095, and 1.024 (ρ); and 1.065, 1.109, and 1.018 (ρ_0)
(9) Model takes into account physical wear and obsolescence $$\dfrac{\dfrac{1}{P}\dfrac{dP}{dt} - \dfrac{T'}{T}}{\dfrac{V}{P} - \dfrac{T'}{T}\dfrac{\overline{K}}{P} - \delta\dfrac{\overline{K}}{P}}$$	0.251	0.233	Depreciation resulting from physical wear and partial obsolescence has already been taken into account in the formula. Depreciation due to inconsistency with structure of demand and

84

Formula	$n_э$	n_{90}	Observations
(10) Model takes into account construction period for capital creation and wear $$\frac{\left(\frac{1}{P}\frac{dP}{dt}-\frac{T'}{T}\right)(1+\beta)^{\mu}}{1-\frac{V}{P}-\delta\frac{\overline{K}}{P}-\frac{T'}{T}\frac{\overline{K}}{P}(1+\beta)^{\mu}}$$	0.30	0.27	disproportions in production remains to be considered. We assume this depreciation $\delta = 0.015$ μ is period of immobilization of capital investment over period of construction of project. It is assumed to be two years, i.e., it is assumed that the project being built begins partial operation only in the third year after start of construction. Then $(1+\beta)^{\mu} = 1.06532 = 1.135; \delta = 0.015$
(11) Model with simultaneous consideration of technological progress and wear $$\frac{\frac{1}{P}\frac{dP}{dt}-\left(\rho+\frac{T'}{T}\right)}{1-\frac{V}{P}-\frac{T'}{T}\frac{\overline{K}}{P}-\delta\frac{\overline{K}}{P}}$$	0.166	0.169	It is assumed that $\rho = 0.014$, $\rho_0 = 0.013$, and $\delta = 0.015$.

which category, as we know, includes only income created in the material sphere of production. Then the right-hand side of the formulas must include capital and labor resources that operate only in the production sphere. However, such a solution provokes doubts because, although nonproduction capital and labor services for workers in the nonproduction sphere do not directly produce material values, they definitely promote their creation by raising the productivity of labor and the use of capital. Therefore it would be incorrect to abstract completely from the influence of these factors. Therefore we give two more calculations: the first includes the totality of basic and circulating capital and all workers in the national economy; and the second includes these same categories, but only for the production sphere (the latter indicators will be indicated by a "zero" subscript).

Statistical data to determine certain parameters which figure in the modifications of the basic model do not exist. We will use approximate values. These parameters include technological progress (ρ) and write-off of capital (δ) due to obsolescence and failure to correspond to the structure of demand.

Now we will calculate the following indicators (see Table 1).

The technological progress parameter ρ may be roughly determined by using the Cobb-Douglas formula. If net output in the initial and following periods is denoted, respectively, by P_1 and P_2, then ρ may be calculated from the following relationship:

$$\frac{P_2}{P_1} = e^\rho \left(\frac{K_2}{K_1} \right)^\alpha \left(\frac{T_2}{T_1} \right)^{1-\alpha}, \tag{11}$$

where K_1, K_2, T_1, and T_2 are capital and labor resources, respectively, in the two successive periods (see Table 2).

Putting the numerical values from Table 1 into the formulas of the preceding section, we calculate the norms of effectiveness. We present them for visual convenience in Table 2, where η_ϑ is the first calculation, which includes all capital and

86

labor resources in the national economy, and η_{30} is the second calculation, which applies to the sphere of material production, i.e., where in the formulas the first column of the indicators T'/T, K/P, and $\overline{K}/P$ have been replaced by T_0'/T_0, K_0/P, and $\overline{K}_0/P$, respectively.

The numerical values of the norm of effectiveness given in Table 2 calculated by formulas (7), (9), and (10) demonstrate the high norm of effectiveness in the USSR, which reaches 25% and higher. In the presence of even modest technological progress (an annual improvement of 1.4% and 1.3% in all), the norm of effectiveness decreases substantially, as the coefficients calculated by formulas (8) and (11) show. But since with technological progress the current cost of output falls noticeably, this condition must lead to a certain increase in η.

The norm of effectiveness when taking into account resources only for the material production sphere (η_{30}), in the absence of technological progress, is somewhat lower or close to the norm η_3, calculated by relating net output to all the resources of the national economy. On the basis of the considerations in the text, it is more correct to orient ourselves according to some intermediate value of η that is between the aforementioned ones.

6. Conclusions

We have shown above that the assumption of instantaneous transformation of capital, i.e., the assumption that capital always can be transformed from one form to another, allowing us to move from one structure of production (labor/capital ratio) to another without loss, represents an extreme abstraction, especially when applied to the socialist economy. (9) The adjustments we introduced into the model have weakened this assumption somewhat. But we must emphasize that even in its original form it is not so strict and unjustified as may seem to be the case at first glance. In a wisely built planned economy, advantages may simply not arise for the transformation of recently created capital, since in the course of planning we anticipate both the need for products and possible technological progress and the relationship between the labor and capital balances. All this must be considered in deciding on the direction for capital investment. Capital is created on the assumption that over the years it will yield the necessary output and will be economically efficient. Briefly put, the hypothesis of instantaneous transformation of capital is applicable because there is no great advantage in such transformations and, therefore, in the use of this assumption.

Finally, we may base some final conclusions on the foregoing calculations. Certain statistical data and parameters must be specified in the future. A methodology must be developed for their determination, especially for the determination of $\overline{K}$, the size of the adjusted capital stock as compared with the acquisition cost due to various causes (physical wear, obsolescence, inconsistency with the structure of demand, and so on).

The important differences between the actual national economy and the abstract model — the multiproduct nature, lack of instantaneous transformation of capital, and insufficient justification of the assumption concerning optimality of decisions in it (not only at a given moment but also in the past) — do not permit the values generated in our calculations to be adopted as a reliable basis to determine the actual value of the norm of effectiveness. But since the aforementioned deviations between the model and reality usually lead to an increase in the norm of effectiveness rather than to a reduction, the data presented to a certain degree contradict the earlier intuitive judgment expressed by Kantorovich about the high value of the norm of effectiveness for the USSR economy — 20-25% and higher.

Considering the great importance of this indicator for planning and economic calculations, we assume that the new approach provided by the model described above may be applied for an approximate, but objective, approach to the calculation of the norm of effectiveness.

Since we do not have other effective methods for precise calculation of the norm of effectiveness for the national economy as a whole, we cannot calculate a fully satisfactory estimate of the precision of our calculation when applying the models. However, such a comparison could be made with other models — one-product and multiproduct models that permit a specific linear-programming calculation. Preliminary calculations show that the analysis of such models may also be applied to specify a methodology for calculating certain parameters $(\overline{K}, \delta,$ and $\beta)$ in our models. These comparisons may be based on analytical calculations of more complex models, such as models of type III (see [7]), which consider the possibility of discarding capital because of its structural inconsistency.

Notes

1) Polish scientists devoted an entire issue of their journal [6] to Fel'dman. Besides giving a translation of his work, this book gives a very detailed mathematical commentary on Fel'dman's model written by V. Psheliaskovskii.

2) The capital K participates in the production process, but it is not expended (the expended part is deducted when calculating net output) and continues into the next period. Thus capital plays the role of catalyst in this production process, influencing labor productivity, and creating output.

3) Total national consumption includes both personal and social consumption.

4) The general form of the Cobb-Douglas function is $P = aF_1{}^{b_1} \cdot F_2{}^{b_2}, \ldots$ $\ldots, F_n{}^{b_n},$ where P is the volume of output; $F_1, F_2, \ldots, F_n$ are different factors of production; $b_1, b_2, \ldots, b_n$ are constant coeffici nts of elasticity of output with respect to the input of the respective factors of production. The total of the elasticities $B = b_1 + b_2 + \ldots + b_n$ may be greater or less than unity. This function is linear in the logs. When $B = 1$ there are constant returns, i.e., a linear homogeneous function. P. H. Douglas and C. W. Cobb, "A Theory of Production," American Economic Review, 1928, vol. 18, supplement.

5) A number of linear programming dynamic models have been calculated at the Institute of Mathematics of the Siberian Section of the USSR Academy of Sciences. They were presented by G. G. Puzanova and L. A. Ponomareva at the spring 1966 Conference on Long-range Planning.

6) This is proved exactly in the article by L. V. Kantorovich and V. L. Makarov [14].

7) $\overline{K}$ is the production value of existing capital, i.e., the value of capital that can ensure (for labor T) the same productivity as the given capital for K. Formula (9) is derived from (7), since the loss of capital stock $\delta \overline{K}(t)$ may be considered formally as consumption.

8) By this we mean the loss of output during the process of assimilation related to the planned annual capacity.

9) In studying a one-product model the assumption of instantaneous transformation of capital is not necessary (see [7]).

Essays in Optimal Planning

References

1) J. M. Keynes, The General Theory of Employment, Interest, and Money, Moscow, Foreign Literature Publishers, 1948.

2) R. Allen, Mathematical Economics, Moscow, Foreign Literature Publishers, 1963.

3) F. A. Fel'dman, "K teorii tempov narodnogo dokhoda," Planovoe khoziaistvo, 1928, nos. 11, 12.

4) E. D. Domar, Essays in the Theory of Economic Growth, New York, 1957.

5) Foundations of Soviet Strategy for Economic Growth, Nicolas Spulber, ed., Bloomington, 1964; 2nd ed., 1965.

6) Studia Economiczne, Warsaw, Polish Academy of Sciences Press, 1963.

7) L. V. Kantorovich and L. I. Gor'kov, "Funktsional'nye uravneniia odnoproduktovoi modeli," Dokl. AN SSSR, 1959, no. 4.

8) L. I. Gor'kov, "Odnoproduktovaia dinamicheskaia model' i analiz ekonomicheskoi effektivnosti kapital'nykh vlozhenii," in the collection Mathematicheskii analiz rasshirennogo vosproizvodstva, Moscow, USSR Academy of Sciences Press, 1962.

9) A. N. Kolmogorov, "Introduction," ibid.

10) L. V. Kantorovich and I. G. Globenko, "Odnoproduktovaia dinamicheskaia model' pri nalichii mgnovennoi prevrashchaemosti fondov," Dokl. AN SSSR, 1967.

11) T. C. Koopmans, "On the Concept of Optimal Economic Growth," Pontificae Academiae Scientiarum Scripta Varia, 1965.

12) D. Gale, Optimal Programs for a Multisector Economics with an Infinite Time Horizon, 1965.

13) Inagaki, Optimal Growth under Technological Progress, Netherlands Economic Institute Publication no. 38/66, 1966.

14) L. V. Kantorovich and V. L. Makarov, "Optimal'nye modeli perspektivnogo planirovaniia," in the collection Primenenie matematiki v ekonomicheskikh issledovaniiakh, vol. 3, Moscow, "Mysl' " Publishers, 1965.

15) Al'b. L. Vainshtein, "Dinamika narodnogo dokhoda SSSR i ego osnovnykh komponentov," Ekonomika i matem. metody, 1967, no. 1.

16) B. N. Mikhalevskii and Iu. P. Solov'ev, "Proizvodstvennaia funktsiia narodnogo khoziaistva SSSR v 1951-1963 gg.," Ekonomika i matem. metody, 1966, no. 6.

9

Once Again on Calculating the Norm of Effectiveness on the Basis of a One-product National Economic Development Model*

L. V. KANTOROVICH and A. L. VAINSHTEIN

The question of the essence and methods of establishing and calculating the capital investment effectiveness norm is one of the most important for the planning of the socialist national economy. Therefore, it was to be expected that the attempt by the present authors to estimate the numerical norm of effectiveness for the Soviet national economy, proceeding from certain theoretical assumptions, on the basis of a one-product model of the national economy and summary national economic indicators [2] would prompt replies and criticisms, including the criticism by one of the founders of the mathematical analysis of the norm of effectiveness (N.E.) — A. L. Lur'e. Since his article contains, along with several correct observations and matters, some objections and assumptions that are incorrect in our view, we consider it necessary to review his criticism and introduce clarity to the discussion of the problem.

As is known, in their earlier work the authors presented the formula for the norm of effectiveness ($\eta_\ni$), i.e., that increase in net output (national income) which is yielded per unit of time

---

*Ekonomika i matematicheskie metody, 1970, no. 3. Reply to A. L. Lur'e [1].

through a properly utilized incremental marginal unit of capital investment. This basic formula (1) establishes the relationship between this indicator and certain global national economic indicators, derived from statistical data [2, p. 703]:

$$\eta_{\mathfrak{I}} = \frac{\partial U(K,T)}{\partial K} = \frac{(1/P)(dP/dt) - (T'/T)}{1 - (V/P) - (T'/T)(K/P)}, \qquad (1)$$

where $P(t)$ is net output, or national income; it is assumed that it is a function of available production capital $K(t)$ (basic and circulating) and labor $T(t)$, i.e., $P(t) = U[K(t), T(t)]$; $(1/P)(dP/dt)$ is the rate of growth of net output; T'/T is the rate of growth of labor; V/P is the share of consumption V in national income; and K/P is the ratio of production capital to national income, or the capital-output ratio.

If, from formula (1), we determine dP, replacing T'/T by the expression dT/Tdt, we get the economically crystal-clear relationship: $dP = \dfrac{P}{T} dT + \eta_{\mathfrak{I}} \left(P\,dt - V\,dt - \dfrac{K}{T} dT \right)$, i.e., the growth of net output is the sum of the increase generated by an increase in the labor force (the first term) and by an increase in capital investment less the consumption required for the growth of the labor force and less the increase in capital required for that labor force increase (the parentheses in the second term).

The following assumptions underlay this formula [2, pp. 700-701]:

1. The function $U(K, T)$ already contains the optimal solution; i.e., we know how to use the available capital and labor (with a given level of technology and organization) in the best possible way, in order to produce the greatest amount of output per unit of time.

2. The function U is positive and homogeneous (of the first degree).

3. The increase in capital stock per unit of time (rate of increase) is equal to net output less the amount going for consumption $V(t)$.

More on the Norm of Effectiveness

4. The hypothesis of instantaneous transformation of capital consists in assuming that the producer always has the desired structure of capital and that this is the optimal structure, since optimization is implied in the very structure of the production function.

The criticism of our work has proceeded in essentially three directions.

1. A. L. Lur'e first assumes that the N.E. must represent the marginal minimal allowable investment effectiveness, and, through the formula in our article (N.E. $\eta_9 = \partial P / \partial K$), he concludes that the "norm of effectiveness turns out to be equal to the marginal value (over time) of the average effectiveness" over the interval dt; consequently, the relative effectiveness of investment in any simultaneously undertaken projects must be identical and equal to the norm [1, p. 371]. Thus the "role of the norm of effectiveness as the lower bound distinguishing between investments that should be undertaken and those that should be avoided disappears."

A. L. Lur'e further believes that the average effectiveness or incremental output-capital ratio $\partial P(t)/\partial K(t)$ must depend on $dK(t)/dt$, since "as more investments are undertaken in a certain period of time and as their intensity $dK(t)/dt$ is greater (in other words, the faster the rate of growth of capital), the relative effectiveness of total investment (the average effectiveness) will be lower." We did not, however, take this into account in our article [2, p. 372].

First of all, the assertion that our N.E. formula, based on the production function $P = U(K, T)$, gives not the marginal but, rather, any effectiveness differing from the average is unjustified. The basic relationship in (1), which gives the N.E., is based on the formula $\eta_9 = \partial U/\partial K$, which, in its content, represents the differential increase in output for an infinitesimally small increase in capital; i.e., it represents the limiting value.

The objection of Lur'e that the average rather than the marginal value is used as the N.E. in our model is wide of the mark, since the N.E., as it is understood in the optimal plan (in the

93

works of L. V. Kantorovich and of A. L. Lur'e himself), is one
of the varieties of objectively determined valuations in the dynamic model (some combination of objectively conditioned valuations of different ingredients). Therefore, it has the same
value for all investments; i.e., in the optimal plan in which objectively conditioned valuation and a single N.E. have any meaning, all investments are made which have an equal, maximum
effectiveness that is, therefore, both the marginal and average,
and is unique. Theoretically — for linear and even multiproduct models — the N.E. is identical for all technologies [3,
p. 293].

Of course, because of deviations from the theoretical linear
dynamic model (integer values, nonlinearity, various constraints,
nonoptimality, and possibilities for more effective investments
resulting from the appearance of new processes and needs),
capital investments are actually conducted over a certain interval of the N.E. but it is, essentially, a small interval; i.e.,
the assumption of an equal N. E. is approximately true in
actual conditions.

As a result, Lur'e's assertion that "during any interval of
time and, therefore, at any given moment of time, projects are
undertaken which have different relative effectiveness — i.e.,
which have different ... incremental output-capital ratios"
[1, p. 371] — does not correspond to reality.

On the other hand, we may expect that in the overwhelming
majority of cases it is rational to undertake investments within
a narrow range of the norm, i.e., the mode will be close to
the mean. The average will not substantially exceed the marginal since highly effective investments in the overwhelming
majority of cases were also efficient earlier and, therefore,
ought to have been undertaken long ago; the less efficient
should remain unrealized for a long time in the future, except
for projects whose construction is necessitated by special considerations (defense or foreign policy). Therefore, in a rational plan the average and marginal valuations, as a first approximation, will coincide.

The application of our formulas to discrete dynamic optimal

models, in which an exact value of the norm of effectiveness is calculated by linear programming methods, may serve as a check on the precision of the calculation of the norm of effectiveness on the basis of a simplified continuous model. If, for example, we assume Kantorovich's simple four-year dynamic model [3, p. 340], with an appropriate calculation of the norm of effectiveness for the first year according to our basic formula (1), the N.E. = 0.371, and for the discrete model we get 0.36.

G. G. Puzanova made certain comparisons for the discrete one-product model that differed from our continuous model (it did not have transformation of capital, and economically obsolescent capital was omitted). The results for discrete and continuous models were very close: in nine out of ten cases the difference between the norms of effectiveness calculated by the objectively determined valuation and by our formula fluctuated between zero and 7.4 percent, and in six of these cases the difference was between zero and 2 percent. (1)

These examples show that in a number of cases the simplified formulas give satisfactory results.

Lur'e sees as another fault in our model its use of a relationship between net output and the intensity of investment (analytically, that P and $\eta_э$ do not depend on dK / dt). (2) We cannot agree with this opinion since P and $\eta_э$ depend on K, and therefore it is not clear that there is a dependence on dK/dt. There is no need to introduce this explicit dependence on dK/dt — i.e., on short-term fluctuations in the share of investment — since the investment norm in reality does not undergo short-term fluctuations that exert a notable influence on the actual effectiveness of investment. This concept would contradict the meaning of the norm of effectiveness as a long-term norm; and its long-term changes have an effect on the size of capital — the capital stock itself — and, therefore, are significantly reflected even in the norm of effectiveness based on our formulas.

In our paper we assumed that total available investment. which represented the difference between net output and consumption, actually was fully transformed into an increase in the capital stock. A number of well-known foreign economists start with this formula (R. Stone, H. Houthakker, and L. Taylor [5]). Lur'e proposes to use the following in place of this formula: $dK / dt = \varphi(\bar{P}(t) - V(t), t)$ [1, p. 373, footnote].

According to this formula, the increase in capital is not equal to available investment but, rather, is only a function of the latter and of time; i.e., available investment is not transformed immediately into an increase in the capital stock; rather, it is undertaken over time and incompletely. It is not clear what the essence, the dimension, of the function φ is which economically characterizes the "interconnection" of capital investments; no information is given on how to derive it. This proposition does not lend itself to a clear interpretation.

2. The second group of Lur'e's observations relates to the assumptions and hypotheses assumed in the model and its modifications. Lur'e objects to our hypotheses about the positive homogeneity of the first order for the national income production function. He presents only the one very general observation that a simultaneous increase of λ in both capital and labor must, as a rule, through the advantages of large-scale production, be accompanied by a greater increase of output [1, p. 375]. But the critic lets slip from his view the notion that in increasing the scale of production, as well as in utilization in this case of less desirable natural resources (which he himself recognizes), transport costs for raw materials and fuel supply, as well as for shipment of finished output, rise considerably, and this leads to a reduction in net output. This circumstance especially reduces the advantage of the large-scale enterprise in agriculture and in the food and light manufacturing industries. The author does not feel that the effect of large-scale production is much lower for the national economy than for individual enterprises in certain industries.

The assumption of a homogeneous production function is a useful and satisfactory approximation in linear models. A non-homogeneous relationship, of course, may take place for individual industries; but, as various Soviet and foreign papers [6, 7, 8] have shown, divergence from it for the national economy as a whole is slight.

In order to estimate the size of the appropriate correction, we will assume that the production function does not have first-degree homogeneity:

$$U(\lambda K, \lambda T) = \lambda^{1+v} U(K, T).$$

Then, repeating the operations performed for the homogeneous model, we get the following formula for η_θ:

$$\eta_\vartheta = \frac{\dfrac{1}{P}\dfrac{dP}{dt} - (1+v)\dfrac{T'}{T}}{1 - \dfrac{V}{P} - \dfrac{T'}{T}\dfrac{K}{P}}.$$

As a result, this adjustment may give some insignificant reduction to η_ϑ, since for the national economy as a whole v is of the order of 0.5. In addition, if we consider the possibility of homogeneity not of the first degree, we get a somewhat smaller value for ρ [2, p. 708, formula (11)] and, as a result, the norm of effectiveness rises. Therefore, this adjustment, together with the adjustment for technological progress, which reduces the value of the former, will not be able to exert any substantial influence in the final calculation.

The estimation of technical progress in our paper was performed by introducing a multiplier $e^{\rho t}$ into the net output production function formula and applying a Cobb-Douglas function to derive a numerical value of ρ.

Lur'e objects to both of these procedures. He feels that the estimation of this factor cannot be based on a "simple introduction (into the production function formula — L. K. and A. V.) of a multiplier that grows constantly over time; the very character and form of the function change." Instead of the national income production function that we used, $P(t) = e^{\rho t} U(K, T)$, he proposes $P(t) = U(K(t), T(t), t)$, in which no consideration is given to the form and character of the relationship between U and t [1, p. 376] , which precludes the possibility for any kind of numerical calculation.

However, linear and exponential (log linear) relationships are the simplest and easiest to interpret because the influence of technological progress in the form of an exponential function, growing constantly through time, represents a simple and acceptable approximation to reality. Foreign and Soviet economists (Samuelson, Solow, Tinbergen, Mikhalevskii, and others) usually use this representation of technological progress. The introduction of innovations and inventions, as a rule, proceeds in stages, extends over a number of years, and thus conditions a stable relative influence on the growth of this exponent. (3)

The technological progress formula that we used may be justified in the following way.

We will take the influence of technological progress over time in the most general form, introducing a dependence upon time into the production function. Then we have: $P(t) = U (K, T, t)$.

Over a certain interval of time near the point t_0 we may approximate the obvious dependence of P on t by an exponential function, choosing ρ such that, for the function

$$U_1(K, T, t) = e^{-\rho(t-t_0)}U(K, T, t),$$

the following relationship holds: $dU_1(K, T, t) / dt = 0$ with $t = t_0; K = K_0; T = T_0$.

In this case U_1 will depend only slightly on t, when it is near t_0, and under the homogeneity assumption even for substantial changes in K and T, if their ratio changes only slightly. Thus we may consider, approximately, that U_1 does not depend upon t, i.e., interchangeably with $U_1(K, T)$ we write:

$$U_1(K, T, t) \approx e^{\rho(t-t_0)}U_1(K, T).$$

Therefore, the assumed hypotheses of technological progress

represents a usually feasible approach to an approximate simplification of the relationship.

And, finally, Lur'e feels that using the Cobb-Douglas formula in its simplest form, $(P = e^{\rho t} K^{\alpha} T^{1-\alpha})$, to represent a national income function, as we did, is generally inappropriate, since equating the sum of the exponents of the factors (K and T) to unity is to assume homogeneity for this function, although, in his opinion, this does not correspond to reality [1, p. 376].

The Cobb-Douglas function as an approximate representation of national income or the nation's output has been used by many well-known foreign econometricians [6, 7, 10, 11] and Soviet economists [8, 12]. In particular, Kornai, in his monograph, which appeared in Hungary at the same time as our work, assumed a national income production function of exactly the same form as we did [7, p. 258], and through it he calculated the influence of technological progress over a 12-year period (1951-62) and obtain satisfactory results.

We used the Cobb-Douglas formula only for an initial estimation of the parameter ρ, which reflects technological progress. But it does not represent a basic assumption in the analysis; rather, it has a supplementary meaning. In general it would be possible to derive ρ from other considerations without recourse to this formula.

Our opponent makes another observation in connection with the Cobb-Douglas formula [1, p. 376, footnote]. This concerns the sharp departure in our article between the effectiveness norms calculated through the Cobb-Douglas formula and through the basic formula (0.13 and 0.22, respectively). In calculating the norm of effectiveness by the Cobb-Douglas formula ([2, formula (7a)] $n_{\scriptscriptstyle 9} = \alpha P / K$), we used the three-factor production function model of B. N. Mikhalevskii and Iu. P. Solov'ev [8, formula (36)]; but in moving from the three-factor model to our two-factor model, the influence of the third factor (natural resources) was distributed proportionally between the other two, although it would have been economically more correct to associate it, as is usually done, with capital. Making this adjustment, we obtain $\alpha = 0.459$, instead of the previously adopted $\alpha = 0.382$ [2, p. 707]. Then η_3 rises from 0.13 to 0.151, which must be considered correct for this case.

The important thing is that the norm of 0.151 reached by the Cobb-Douglas formula must be compared not with the norm of 0.22 calculated by the formula [2, p. 707, formula (7)], as was incorrectly done by Lur'e, but with the norms calculated by us through the basic formula, taking technological progress into account [2, formulas (8) and (11)], since the Mikhalevskii-Solov'ev equation also was made to take account of technological progress. (4) Formulas (8) and (11) [2] give a norm of effectiveness equal to 0.146 and 0.166, respectively, and fully correspond to the corrected norm of 0.151 derived through the Cobb-Douglas formula. Nor do they differ substantially from the previously calculated 0.13.

3. The third kind of criticism relates to the numerical results that we calculated. Lur'e feels [1, pp. 370 and 375, footnote], that the norm of 20-25 percent, which was intuitively given earlier by Kantorovich, is not supported by the estimates of the authors of the model and represents an exaggeration.

The establishment of even an approximate value for the norm of effectiveness is of the greatest practical importance. Therefore, we have taken the formulas presented in our paper (which were obtained by considering various factors) and from them we have constructed the relatively most reliable variants, taking account of all the positive and negative adjustments: to wit, a combination of formulas (8) and (10) [2, p. 705]:

$$\eta_3 = \left[\left(\frac{1}{P}\frac{dP}{dt} - \frac{T'}{T} - \rho \right)(1+\beta)^\mu \right] \Big/ \left[1 - \frac{V}{P} - \delta\frac{\overline{K}}{P} - \frac{T'}{T}\frac{\overline{K}}{P}(1+\beta)^\mu \right].$$

From here, putting in the corresponding numerical data from Tables 1 and 2 [2], we get $\eta_3 = 0.20$.

It should be noted that this calculation of the norm of effectiveness envisages a normal two-year lag, and we should assume an immobilization of capital only when it exceeds this period.

Notwithstanding the obvious tentativeness of the calculations, the inadequacy of data, and the incomplete correspondence between the real economy and the hypotheses of the model, we nevertheless feel that it is possible to adopt the value 0.2, which we have found, as an approximate value for the norm of effectiveness.

This does not contradict the previously given intuitive opinion of a norm of effectiveness in the range of 20-25%, since until now the potential volume of capital investment has increased significantly and large new projects having great efficiency have gradually been realized, as a result of which the norm of effectiveness in recent years, naturally, may have fallen considerably. The significant technological progress of the last few years has operated in the same direction (compare the formulas taking account of and excluding technological progress in our work [2, p. 707]).

We take $\eta_9 = 0.2$ as an approximate value of the norm of effectiveness and believe that the norm of effectiveness, in fact, should be slightly higher for the following reasons.

The model assumes optimal utilization of the capital stock in the real economy; this does not in fact exist. Using multiple products in the model will lead to a great interconnection of investment and a reduction in free responses which must raise the norm of effectiveness. Elimination of the effects of nonoptimality — the existence of nonoptimal decisions, irrationalities, and disproportions — opens up new possibilities for extremely efficient use of investments in the course of transition to an optimal policy and, in conjunction with this, will also lead to a higher value of the norm of effectiveness.

In moving from a nonoptimal to an optimal regime, the rate of growth of national income will increase. According to our formula (1) for the norm η_9, this should also increase the norm of effectiveness.

The aforementioned adjustments, which tend to increase the norm of effectiveness and which must be made for an optimal economy, evidently exceed considerably the difference between the average norm of effectiveness, calculated by our formula, and the norm of effectiveness advocated by Lur'e; and, as a result of these corrections, the norm of effectiveness will probably turn out to be higher than the one we calculated earlier. This is also indirectly acknowledged by Lur'e, who writes [1, p. 375]:

" ... we should mention that the statistical data employed characterize a process of growth of the national economy which does not conform to the requirements of optimality, ... which could not but reduce the total effect of capital investment. Therefore, there is no certainty that the ratio $\partial P(t) / \partial K(t)$ (i.e., our norm of effectiveness — L. K. and A. V.) would not be considerably higher if the development of the national economy corresponded more closely to the prerequisites of the model of optimal economic decisions."

We made some controlled calculations of the norm of effectiveness according to our formula (1), on average, for the two years (1966-67), although not for the entire national economy but, rather, for a group of only two sectors: industry and construction in actual prices. The norm of effectiveness was calculated to equal 0.243 = 24.3 percent, as against the 22 percent that we presented earlier for the national economy as a whole [2, p. 707, formula (7)].

Thus the actual norm of effectiveness, on average, for the last two years for a set of two sectors that account for more than half of the capital investment and provide more than half of the national income, calculated by our basic formula, supports the correctness of our earlier estimate for the economy as a whole and, therefore, the correctness of the minimum $\eta_0 = 0.2$ which we have adopted and advocated.

* * *

We will present our results. Lur'e's article, which is interesting methodologically, presents unjustifiably stringent demands on the mathematical model necessary for analysis and rough calculation of the norm of effectiveness. However, our paper in many respects has the character of an extremely preliminary investigation; it is based on simplified assumptions and makes only the most approximate conclusions, which our opponent has not considered to a sufficient degree.

The history of science shows that scientific research, especially in its assumptions, has frequently been based on heroically simplified hypotheses in the absence of adequate information, incomplete analysis, and only partial correspondence between the scientific description and the phenomenon studied.

Technological calculations of durability and calculations of the flow of liquids had been based for 200 years on the simplified one-dimension model originating with Euler and Bernoulli; only with the appearance of electronic computers have we begun to use exact systems of equations from the theories of elasticity and hydrodynamics. The same situation has been observed in many areas of technological science. However, practice has shown the possibility and adequacy of also using simplified mathematical models in the study of and decisions on concrete economic problems when there are inadequate indicators and incomplete information.

Lur'e, it is true, acknowledges in principle that "it would be incorrect to assert that a simplified continuous dynamic model, determined by the conditions given earlier in the (Kantorovich-Vainshtein) article, and its modifications cannot serve as a weapon for theoretical analysis of certain economic problems. But the conclusions from these models should be approached with great caution."

However, the authors of the paper under criticism have shown great care and discretion in presenting its results. The suggestions made in it are considered extremely preliminary and are accompanied by a number of reservations and a variety of formulas. In conclusion, the authors wrote [2, p. 709]: "Of course, it would be premature to base any kind of final conclusion on the calculations which we have made. Certain statistical data and parameters require further specification. A methodology for determining them must be developed...."

A number of second-degree factors and small adjustments

might also be taken into account and introduced into the model when it is developed further. This does not discredit its basic ideas.

The roughness of the description given by macroeconomic models is softened, to a certain degree, by the fact that these models are used for global statistical indicators of the economy that are rather stable and, thanks to mutually offsetting deviations, more reliable. The parameters entering the model (rate of growth of national income, labor, and capital and the share of consumption in the national income) are statistically determinate and are known to every economist. At the same time, the corrections and proposals of Lur'e are either unclear in their construction and function or may not in general be filled out with a numerical content even if they did present certain advantages over our formulas in their mathematical aspect. Incidentally, our opponent nowhere attempts to give formulas for calculation; on the contrary, his demonstration of the supposed lack of substance of analytical calculations of the norm of effectiveness is an exercise in mathematical complication. The authors also see a main virtue of their model in the way in which it permits a real calculation of the norm of effectiveness; this would be impossible for more complex models in the present conditions of inadequacy of published statistical theories. The formulas we have presented are valuable because they give the order of magnitude of the norm of effectiveness and they have already been used to calculate the norm of effectiveness; the results they have provided have been used in the GDR, Poland, and other countries [13]. Besides practical application, the proposed approach has, there is no doubt, a methodological significance. The authors' model makes possible quantitative investigation of the analysis of various factors on the norm of effectiveness — lags, technological progress, depreciation and obsolescence, etc. — and through this, other indicators.

What does Lur'e propose in place of our model? A method of trial and error: "... considering the planned volume of capital investment, the branch structure of the national economy,

and the growth of the labor force, a more or less probable
value of the norm of effectiveness is applied (where do we
get it from? — L. K. and A. V.). Proceeding from this value,
a national economic plan variant is calculated ... and we determine whether it requires greater or lesser investment in
comparison with actual possibilities. In case of 'shortage' of
the assumed investment fund, the effectiveness norm is increased, and in case of 'surplus' it is reduced. Several (sic!)
such iterations will give an approximate (emphasis ours —
L. K. and A. V.) value for the form, which must be corrected
afterwards" [1, pp. 367-68].

Thus Lur'e's approach requires the numerical calculation
of many variants of a detailed national economic plan for a
number of values of the norm of effectiveness in order to reach
a still extremely approximate value for this coefficient — a
grandiose procedure in labor intensity whose execution is completely unrealistic, at least in the next few years.

On the other hand, the simplified approach of the present
authors has been based on a logically built and rather general
system of models. In spite of its imperfection and the statistical difficulties in executing it, it is — for the time being —
one of the few truly accessible and, to a definite extent, scientifically justified approximating approaches to the calculation of
the norm of effectiveness; therefore, in spite of Lur'e's opinions,
it may be offered for further development and application.

Notes

1) In [4] Puzanova shows the following ratios between the norm of effectiveness, calculated through the objectively determined valuations, and the norm calculated by our formula: for the variant without technological progress ($\rho = 0$) it
was 0.996, and for the variant with technological progress ($\nu = 0.02$) it was 1.021.

L. A. Ponomareva has made similar comparisons for multiproduct models
based on her data in [4] for three years. The results, working with objectively
determined valuations, gave a norm of effectiveness of 26-41 percent, and the
results for the continuous model gave a norm of 32-41 percent, i.e., they turned
out to be quite close.

2) According to Lur'e, the N.E. should "represent the limiting ratio of the
increase in intensity of output to the increase in intensity of investment," and

105

he gives a mathematical formalization for this definition [1, pp. 373-74, formulas (7a) and (6)]:

$$\eta_\vartheta = \partial \frac{dP(t)}{dt} \Big/ \partial \frac{dK(t)}{dt}, \text{ where } \frac{dP(t)}{dt} = f\left[\frac{dK(t)}{dt}, K(t), T(t)\right].$$

However, this definition is not elaborated and the model it depends on is not indicated. It is also not completely clear where we get the function f and how we derive the N.E. Incidentally, it is not clear for Lur'e himself (see his footnote). This definition of the norm of effectiveness differs radically from that used earlier [1, p. 366], where the author gives a formula for reducing the expenditures of any kind of resource during the preceding time period t to terms of the same resource expended at time t_1: $v_{np} = v(1+E)^{t_1-t}$, where E is the norm of effectiveness. The relationship between these two definitions is not explained by the author.

3) Denison, studying the influence of "advances in knowledge" on the growth of the national income, reached the conclusion that during the 1950-62 period (and earlier, in the 1925-50 period) there was an "approximately constant rate of advance in knowledge and a small change in the lag" [9, ch. 20, p. 282]. Under the heading "advance in knowledge," this author includes not only technology but also improvements in organization and managerial knowledge. By lag Denison means the interval of time between the appearance of new knowledge (discoveries, inventions, etc.) and the application of this knowledge in practice.

4) In our paper, it is true, this was not stipulated; but Lur'e should have seen this himself if he had turned to the original source we cited. The last term in this equation immediately shows the extent of the influence of technological progress.

References

1. A. L. Lur'e, "Concerning Calculations of the Norm of Effectiveness and a One-product Continuous Model of the National Economy," Ekonomika i matematicheskie metody, 1969, vol. V, no. 3. [Translated in Matekon, 1969, vol. VI, no. 2, pp. 196-216.]

2. L. V. Kantorovich and A. L. Vainshtein, "On Calculating the Norm of Effectiveness Through a One-product Model of the Economy," Ekonomika i matematicheskie metody, 1967, vol. III, no. 5.

3. L. V. Kantorovich, Economic Analysis of Optimal Resource Use, Moscow, USSR Academy of Sciences Press, 1959.

4. G. G. Puzanova, "Some Experimental Calculations Through a One-product Dynamic Model"; L. A. Ponomareva, "Construction and Calculation of a Simplified Dynamic Model of National Economic Planning Based on Information in the Input-Output Table," in Optimal Planning, no. 8, Novosibirsk, "Nauka" Publishers, 1967.

5. A. L. Vainshtein, "A Mathematical Model of Consumer Money Flows," in Optimal Planning and Improvement of Control of the National Economy, Moscow, "Nauka" Publishers, 1969.

More on the Norm of Effectiveness

6. J. Tinbergen and H. Bos, Mathematical Models of Economic Growth, Moscow, "Nauka" Publishers, 1969.

7. J. Kornai, Mathematical Planning of Structural Decisions, Budapest, 1967.

8. B. N. Mikhalevskii and Iu. P. Solov'ev, "A Production Function of the National Economy of the USSR, 1951-1963," Ekonomika i matematicheskie metody, 1966, vol. II, no. 6.

9. E. Denison, Why Growth Rates Differ, Washington, 1967.

10. G. Tintner, Introduction to Econometrics, Moscow, "Statistika" Publishers, 1965.

11. R. Solow, "Technical Change and the Aggregate Production Function," The Review of Economics and Statistics, 1957, vol. 39, no. 3.

12. V. A. Trapeznikov, "Problems of Control of Economic Systems," Avtomatika i telemekhanika, 1969, no. 1.

13. W. Przelaskowski, "Współczynnik sprawności gospodarki narodowej," Ekonomika i organizacja, Czerwiec, 1969.

10

Estimating the Effectiveness
of Capital Expenditures*

L. V. KANTOROVICH, V. N. BOGACHEV,
and V. L. MAKAROV

In the course of technological planning of individual construction projects, when determing which design and planning variants to use, it is always necessary to compare the one-time
expenditures (capital expenditures) and current costs (operating
costs) to find the solution which gives the most profitable combination of the various kinds of social expenditures. The typical
character of economic problems, in which reduction of operating
costs by an increase in capital expenditures is attempted, has
quite understandably led to an attempt by planners and designers
to find principles which define a rational combination of these
expenditures. As early as the 1920s, the problem of "effectiveness of capital investment" was stated with full clarity [1]. In
the 1930s, engineers and designers, working primarily on energy and railroad transportation (see, for example, [2, 3]),
formulated the idea of the norm of effectiveness — a planning
standard expressing the social value of the resources in a capital investment and at the same time defining the desirable
levels of their expenditure in each particulare case.

Later analysis [4-7] uncovered a basic connection between
the problem of "effectiveness of capital investment" and the

*Ekonomika i matematicheskie metody, 1970, no. 6.

general idea of optimal planning and operation of the socialist economy. The development of mathematical programming created the possibility of interpreting the traditional problem of the selection of design alternatives in the clear terms of optimal linear (or, more generally, convex) programs [8]. It does not follow from this that studies of the optimal utilization of capital resources have only a purely historical significance as one of the sources of the formulation of optimal methodology in our national economic literature. Although the problem of distributing a given amount of investment among various desirable projects represents a partial (and, in some situations, an extremely simplified) case of the general problem of rational utilization of scarce resources, optimal problems subject to capital constraints have a special significance which is conditioned by the economic meaning of the dual capital valuations. The norm of effectiveness of capital expenditures occupies a special place in the system of optimally conditioned valuations which determine the total economic significance of utilizing production resources and at the same time the economic value of time as an important good (see [4-9, 10]). Because of the economic meaning of the optimal valuation of capital expenditures and its importance in economic growth, the problem of "effectiveness of capital investment" in production processes (more precisely, the problem of the numerical determination of the correct interpretation and adequate application of the norm of effectiveness) acquires an independent and, to a certain degree, autonomous significance in the whole complex of problems of optimal planning. (1)

The universal character and fundamental economic significance of the norm of effectiveness of capital investment, as well as the rich traditions of its theoretical analysis in Soviet economic literature, led to the application of this category of optimality analysis in practical economic calculations long before the optimality approach to problems of an economic character received wide recognition. The "Standard Methodology for Determining the Economic Effectiveness of Capital Investment in New Technology" [11], published in 1959, following extensive discussion which took place in an all-union conference, legitimated the application of the norm of effectiveness E in all cases for the selection of capital investment alternatives and recommended the formula for "expenditure in commensurable terms" (C + EK) as the critical indicator which would ensure correspondence between local decisions and rational utilization of investment on a national economic scale.

It is easy to see the connection between the basic recommendations of the "Standard Methodology" and the various kinds of optimal programming. This can be seen immediately since the value of E that was introduced was calculated to be far greater than the depreciation norm — the value of E was determined to be in the range of 0.1 to 0.33 — and at the same time was not interpreted within the framework of traditional economic calculations. At the same time, it is natural to use it as the optimal valuation for limited capital resources in extremal problems for minimization of direct costs (calculated according to standard accounting practice) or for maximization of income (under certain additional assumptions). The norm E is also recommended by the "Standard Methodology" for discounting to comparable terms different expenditure and income time streams. In this case the norm E cannot be interpreted on the basis of traditional ideas; it clearly does not coincide with the rate of growth of labor productivity, or with the rate of increase of capital stock, or with any other kind of economic growth indicator used in traditional practice. Finally, the commensurated sum C + EK represents, as is easily seen, the simplest form of differential growth-related expenditures, which, certainly, is justified only under the observance of a number of special conditions which abstract significantly from economic reality.

The scientific achievements of the decade now ending and the immeasurably greater demands on applications (as well as the potential for them) inexorably require methodological developments which shed light on the basic questions of the choice of investment alternatives and the evaluation of their effectiveness at the current level of technology and advanced experience. In our opinion the recently published second edition of the "Standard Methodology" [12] does not reflect, to the required extent, the present state of the problem and does not represent substantial progress over the first edition.

In our article we attempt to some extent to fill this gap and to present methodological principles relating to the effectiveness of capital investment. The article has primarily a positive character; its arrangement is determined by systematic consideration of the essence of the problem of evaluating the effectiveness of capital construction projects (not the structure of the "Standard Methodology"). Primary attention is devoted to the general theoretical conditions that come out of the theory of optimal programming; these conditions, and the calculating formulas which correspond to them, are in our opinion an absolutely essential element for the methodology of capital investment effectiveness intended for widespread practical application. Some matters that have been unresolved and insufficiently elaborated to date are noted in passing and are discussed to some extent.

We assume that these principles can be applied both in future methodological elaborations, such as industry methodologies, and in the estimation of the effectiveness of specific projects in actual conditions, which cannot always be satisfactorily done on the basis of the adopted methodology.

The utilization of exact methods for analysis is especially necessary for estimating the effectiveness of measures for technological progress in conditions of a rapidly developing economy, for the estimation of the effectiveness of the production of new products, and for the introduction of new kinds of raw materials and new technological processes.

The idea and purpose of methodological recommendations for evaluating the effectiveness of capital investment consists, obviously, in proposing approaches and methods for analysis of capital investment projects which will reflect, in local calculations conducted on the basis of information that exists in individual units of the economic system, the full national economic effects that would result from undertaking the capital investment being analyzed. In other words, the "Standard

Methodology" must recommend a selection criterion which ensures that the decisions made locally correspond to the interests of the national economy. If it does not do this it serves no purpose.

This central methodological idea gives rise immediately to a number of questions. In general, is it possible to evaluate the desirability of a specific capital investment without reviewing the national economic plan as a whole and utilizing only information contained within a project? Do there exist any approximating methods for the estimation of the effectiveness of projects which may be applied if some of the conditions that are supposed to ensure that local decisions correspond exactly to the national economic plan according to a single criterion are violated?

Without going into the details of the history of the problem, we will note that modern models of optimal planning and control contain the answers to the foregoing questions and the formulation of conditions under which these answers are justified.

1. Some Consequences of the Theory of Optimal Programming

The central problem of optimal programming is the search for conditions (necessary and sufficient) which determine whether a system is or is not in an optimal state. For practical applications it is extremely desirable that these conditions be rather easy to verify. The solution of this problem for convex programming is well known. It is given by the theories of duality (or, in other terminology, by the theorems on the characteristics of optimal decisions). We may point to the well-known analogy between the previously mentioned problem of optimal programming and the problem of estimating the national-economic consequences of a capital investment project which we discussed above. In the first case we are talking about indications of optimality of an economic system as a whole, and in the second about methods to determine whether the national economic plan as a whole will improve through the execution of a given capital construction project. We are talking about more than superficial analogies in this case; relying on the properties of optimal solutions formulated by the general theory of convex programming, it is possible to show ways to estimate the effectiveness of capital investment projects.

We will mention here some results from the theory of mathematical programming which will be necessary for what follows.

We will denote by $X \subset R^n$ the set of the production possibilities in the economic system; $x \in X$ will be a production process described by the vector $x = (x_1, x_2, \ldots, x_n)$, where x_i is the amount of product i produced or expended (per unit of time) in the production process x. In an extremely general form the extremal planning problem may be formulated in the following way: given a set of X and a vector of constraints $b = (b_1, \ldots, b_{n-1})$, find the plan $x = (b, \mu) \in X$, such that $\bar\mu = \max_{(b,\mu)\in X} \mu$. If X is an arbitrary closed convex cone, this extremal problem is equivalent to the general convex programming problem.

<u>Theorem on the characteristics of the optimal solution of the convex programming problem</u> (see, for example, [13]). Suppose we have a convex programming problem $\{X, b\}$ with the following properties: (1) the negative orthant [ortant] of the space of R^n belongs to X; (2) either there exists μ such that (b, μ) is an interior point of the set X, or else X is a polyhedral set; (3) there exist μ' such that $(b, \mu') \notin X$. In order that $\bar x$ be an optimal solution it is necessary and sufficient that there exist a system of prices $\pi = (\pi_1, \ldots, \pi_n)$ such that: 1. $\pi \geqslant 0$; 2. $\pi_n = 1$; 3. $\pi x \leqslant 0$ for all $x \in X$; 4. $\pi\bar x = 0$.

If the closed cone X is not convex, properties 1-4 of the system are sufficient but are not necessary. Optimal solutions of the nonconvex programming problem cannot be described so exhaustively and simply through the properties of a system of prices.

From the foregoing theorem it follows that, if there arises (is discovered, invented, or planned) a new production process $\tilde x \in X$ which is not used in plan $\bar x$, the desirability of including it in the plan is determined by comparing the inner product $\pi\tilde x$

with 0. If $\pi\tilde{x} < 0$, then there is an unconditional possibility of improving the optimal solution $\bar{\mu}$ by means of the new production process. If $\pi\tilde{x} > 0$, the new method is generally applicable (2), although a number of questions which are important in practice remain unanswered (for example, the volume to be introduced or the intensity of utilization of the new process).

The essence of the methods described below for the evaluation of capital investment projects consists in the representation of each product in the form of a "process," an evaluation of the inputs and outputs, and an algebraic summation of all of these in addition to a comparison of the results with zero. The very same principle (perhaps unconsciously) underlies the recommendations of the "Standard Methodology." We should also emphasize that the practical application of this principle runs into a number of difficulties. The major ones are: (1) the condition $\pi\tilde{x} > 0$ does not say anything about the optimal intensity of utilization of process $\tilde{x}$; (2) the new method $\tilde{x}$ may include inputs and outputs which do not enter the plan $\bar{x}$ and which, as it turns out, do not have assigned values; the expression $\pi\tilde{x}$ in this case loses its meaning; (3) the initial plan actually given may not satisfy the condition $\bar{\mu} = \max \mu$, and the prices utilized in it which are used in the calculation of $\pi\tilde{x}$ may depart seriously from the optimal prices which meet conditions 1-4 of the theorem stated above.

2. The General Formula for Calculating the Absolute Effectiveness of Capital Investment

Difficulties in the calculation of the absolute effect of a capital investment in practice are encountered in various combinations, but in most cases they may be avoided. To begin with we will consider a scheme for calculating the effects of investments "in pure form."

What is the economic meaning of the value πx for an arbitrary process x? It is well-known [14] that in models of optimal planning we mean by "ingredients" kinds of products, services, labor, natural resources, and capital assigned to a definite place and interval of time of operation. The vector $x = (x_1,$ $\ldots, x_n)$ includes all inputs and outputs produced with a definite time stream and location. Consequently, the value $\pi x = \pi_1 x_1 + \pi_2 x_2 + \ldots + \pi_n x_n$ represents the pure effect (profit) from the utilization of processes x and prices π throughout the entire time of its operation. Different methods for determining the economic effectiveness of capital investments, as will be shown below, consist of different methods of calculating or valuating the values of this pure effect.

Any capital investment may be represented in the form of a production process. The components of the vector $x = (x_1,$ $\ldots, x_n)$, which describes the capital investment, will represent all possible ingredients of construction expenditures: building materials, the use of construction machinery, different kinds of labor, engineers and designers, and so on. However, in practice it is not necessary for the calculation of πx to represent the capital investment in the form of a vector x consisting of such fractional input components. The capital investment process may be described by a vector of much smaller size, the components of which are the partial sums $\sum_i x_i p_i(t)$, which include all the expenditures over a given interval of time (in practice, one year of construction), aggregated according to prices $p_i(t)$. These prices, naturally, must relate to the period in which the expenditures being aggregated are made. In practice, construction estimates (as well as the future income of the enterprise being built) are calculated without taking account of the time schedules for the expenditures, so that, for example, one thousand bricks are valued the same whether they are used in the first or fifth year of construction. We cannot criticize existing practice for this since we do not usually have at our disposal any other prices beyond those which exist at the moment of calculation; but out of this difference between the demands of strict theory and the actual possibilities of practice there flows a specific methodology for measuring the effectiveness of "capital investment" which we will discuss further below.

Thus the capital investment is a vector of the form

$$(-K_1, -K_2, \ldots, -K_\lambda, 1 \ldots), \qquad (1)$$

where $K_t = \sum x_i p_i (t)$ is the total of all the input ingredients

in the creation of a construction object which are employed in period (year) t and aggregated according to prices $p(t)$, reflecting the conditions of the plan of the t-th period; λ is the period of construction of the project, and 1 symbolizes the fact of completion of the object. The production object — the result of the construction expenditures — represents a new ingredient of the plan; as such it has no price for a single one of the periods preceding its emplacement. Since calculations of the effectiveness of capital investment have meaning only as a way of determining the desirability of undertaking projects, they necessarily must precede the start of construction, i.e., they must be performed at a time before the object can be economically valued. The total of the first λ components of the vector (1) describes only the construction expenditures, but vector (1) does not contain the components which express the national-economic value of the object being designed which must be compared with the expenditures.

The aforementioned difficulty may be resolved easily if we consider that the valuation of any element of the social production apparatus is determined by the conditions and results of its operation in the optimal plan. The production or operation of the facility which is created may be described by a vector of the form:

$$(\ldots, -1, \Pi_{\lambda+1}, -C_{\lambda+1}, \Pi_{\lambda+2} - C_{\lambda+2}, \ldots, \Pi_{\lambda+T} - C_{\lambda+T}, \Pi_{\text{осt}}), \qquad (2)$$

where the negative components reflect the fact of productive utilization of the object which is created by process (1), and Π_t and C_t represent the corresponding aggregates of output and operating expenditures over the corresponding time intervals, measured according to prices $p(t)$, which describe the economic environment of all periods in which the operating expenditures are made or the production produced by the newly created facilities. Combining vectors (1) and (2) gives a complex "pro-

116

cess of production and operation" for the facility, and is described by the vector:

$$(-K_1, \ldots, K_\lambda, \ldots, \Pi_{\lambda+1} - C_{\lambda+1}, \ldots, \Pi_{\lambda+T} - C_{\lambda+T}, \Pi_{\text{ост}}), \qquad (3)$$

in which the object of the capital investment itself appears as an "intermediate product" with a 0 value for external production (expenditures). In vectors (2) and (3) T is the period of operation of the facility being projected and $\Pi_{\text{ост}}$ is its residual value in prices of period $\lambda + T$.

Adding the components of (3) is simply the calculation of the inner product $\pi \tilde{x}$ previously described by the theory as a reliable and objective way to evaluate the desirability of including the process $\tilde{x}$ in the plan (in the case of the capital investment project being analyzed).

Of course, identity of these two calculating procedures depends upon the precondition that the dynamic system $p(t)$, which is used to construct the aggregated components of vector (3), possesses the properties of optimal valuation of the dynamic model of the plan. In actuality, as was noted, practical applications can make use only of a system of prices for some definite period, and these prices may be far from optimal. We will assume that the prices of the period when the calculation is conducted satisfy the necessary requirements, or that there exist ways to correct the existing system of prices in the direction of the system of optimal valuations. Then there remains the problem of moving from the static system of valuation given at the moment of calculation to prices which may be used to measure future expenditures (capital and operating) and the future production of the projects being considered. It will also contain the specifics of the methodology for determining the effectiveness of capital investment.

One of the basic properties of optimal prices is a decline in the general level over the course of time [14-16]. No comparisons of expenditure and income, assigned to the different periods, are possible without taking account of this tendency. Common sense and economic intuition long ago anticipated this assumption of optimal programming: in actual economic calculations it has long been customary to consider equally great "physical" volume expenditures having different time streams to be of "different value," with the norm for discounting expenditures over time usually taken (as we will see, not without justification) as the norm of effectiveness E — the same norm as was used to commensurate capital expenditures and constantly recurring operating expenditures.

Let E be the annual rate of decline of the general price level in the dynamic optimal plan. Of course, this very value is variable over time, but the substitution of the actual rate of decline of the general level of optimal prices (which varies from year to year) by some average value will introduce an error into the calculation that hardly exceeds the possible forecasting errors for future technological progress and other dynamic economic factors, the latter necessarily being taken into account in setting up the optimal national economic plan for a long time ahead. If the components of vector (3) are figured in prices of the period in which the calculation takes place, it is necessary, when adding them up, first to "discount them" to prices of the appropriate periods. This is done by multiplying each component relating to the t-th year by the value $(1 + E)^{-t}$. Under these conditions, the general formula for calculating the effect of production process $Э = \pi \tilde{x}$ is transformed into the formula:

$$Э = \sum_{t=0}^{T+\lambda} (1 + E)^{-t} (\Pi_t - C_t - K_t + D_t - R_t) + \Pi_{\text{ост}} (1 + E)^{-(T+\lambda)}. \quad (4)$$

Besides the symbols used in (3), we use here R_t and D_t. If the expenditures are calculated, as is done in planning practice today, C_t and K_t do not include social cost rental components: national-economic costs of utilization (rental payments) or exhaustion (capitalized rent) of nonreproducible natural resources. R_t expresses this social cost rental component. By D_t is understood "the external effect" of the project which is not reflected in the earnings from the sales of products produced by the new facility. These external effects include: (a) a change (improvement or deterioration) in labor conditions;

(b) a change (facilitation or complication) in the opportunities to undertake other capital investments (for example, the creation of a construction base, access roads, and other infrastructure elements which change the conditions and further construction in the region or locale, and so on); (c) the effect of utilizing the output by consumers (the difference between the selling price, calculated as the income of the units being analyzed which is equal to the effect from utilizing a marginal unit in consumption, and the real effect from utilizing the units of output which precede the "last" or marginal unit); and so on. Obviously, these rental costs, and in particular the external effects (positive and negative), lend themselves to numerical determination only approximately and with great difficulty. These difficulties may be considerably lightened if we describe capital investments with the help of "expanded process vectors" in which the external effect of capital investments appear as intermediate products of the production process being analyzed.

Thus the effect of any capital investment project may be expressed through measurement of all expenditure components (direct and indirect) and output, applying to the initial system of prices a generalizing norm of price decline over time $(1 + E)^{-1}$. We should, however, observe that in a dynamic system of optimal prices the valuations of individual ingredients of the plan decrease at varying rates. For example, the values of railroad shipments, serial production of machinery, and production of a number of new products fall more rapidly than the average rate of decline, while the values of labor and certain raw materials decline more slowly (for details see [15]). Therefore, in discounting costs and benefits, calculated in prices of the initial moment of the calculating period, to the ultimate point which takes account of price movements, it is not enough to utilize an average multiplier, but, rather, it is desirable when necessary to introduce adjustments for the variable movement of prices of different kinds of products and expenditure components. It is more convenient to utilize these adjustments in the calculation of C_t and K_t if sufficient basis and necessary prerequisites exist for each.

Differentiation of the dynamics of values has been, obviously, one of the unrecognized reasons for requiring differentiation of the norm of effectiveness by the industrial branch. Since the dynamics of specific prices are not identical for all products and production costs of the industries, it is not desirable to take account of these differences in the industry norms. An appropriate calculation of the cost and output components is an adequate way to reflect the differences in price dynamics in calculations.

Thus (4) is the most general formula for determining the economic effect of capital investments and conforms completely to the principle of comparing πx to zero, providing, of course, the prices used in the calculation of the components of (4) are optimal, if only approximately, for the initial year of the dynamic plan, and the dynamics of the optimal system of prices are rather well approximated by the average annual rate of decline $(1 + E)^{-1}$.

3. Simplification of the General Formula for Practical Application

By introducing a number of simplifying assumptions in (4), we get several specific cases which are convenient for practical utilization. This affects, first of all, the well-known minimization criterion $C + EK$, the only specific calculating formula from the "Standard Methodology" of 1959. In the second edition calculation of costs by $C + EK$ remains the only calculating scheme for analyzing capital construction alternatives. Other formulas introduced in this document are intended for calculation of the profitability of already existing capital after the fact.

We observe, first of all, that the criterion $C + EK$ (5) is a specific method for evaluating the so-called "relative" effectiveness of the expenditure alternatives. In this special formulation of the problem of measuring the investment effectiveness, it is assumed that the alternatives which can be compared according to (5) give the same national economic effect regarding the volume and composition of basic and "supplementary" production,

The Effectiveness of Capital Expenditures

time streams, and so on. The positive components of vectors of type (3) under these assumptions do not differ from one vector to the next (the rule of "identity of national economic effect"), and differences between the alternatives being compared consist only in the size of the negative expenditure components of (3). Then the basic formula (4) becomes:

$$\Im_{cp} = \sum_{t=0}^{\lambda+T} (1 + E)^{-t} (-C_t - K_t - R_t), \qquad (6)$$

and, naturally, that alternative is preferred (from the entire set whose elements are identical in their national economic effect) with the greatest value of $\Im_{cp}$.

The discounted costs of (5) are a special case of (6) for the following conditions:

a) the service life of the variants being evaluated is infinite ($T = \infty$);

b) the structure and level of production costs in the alternatives being evaluated by (5) are constant over time (and the prices for all cost components have the same dynamics);

c) the capital investment is executed in a single year which immediately precedes the start-up of the facility;

d) in setting up and operating the facility, no nonreproducible resources are used (there are no rent components). Fulfillment of conditions (a)–(d) means that $\lambda = 1$, $K_t = K$, $C_t = C$, $R_t = 0$, $T = \infty$, and instead of (6) we have

$$\Im_{cp} = - C \sum_{t=0}^{\infty} (1 + E)^{-t} - K = - \frac{C}{E} - K. \qquad (7)$$

Maximization of this sum, obviously, is equivalent to minimization of discounted costs.

We must note the artificiality and extreme severity of the conditions assumed by the method of "relative" effectiveness. In practice, when making technical-economic calculations, there are too many cases when it is necessary to choose among alternatives which differ in the volume and composition of out-

121

put (or production time streams). The method of "relative" effectiveness in these cases, strictly speaking, is inapplicable. In addition, planning and preplanning project-making frequently must deal with alternative trends in investments which differ so strongly in character and in the size of their national-economic effect that to reduce them to conditions of "identical effect" represents, in truth, an insoluble puzzle. It seems to us that, in general, the measurement of the total or "absolute" effect of investments must be considered as the basic method for analyzing capital investment projects, reserving for the "relative" effectiveness a subsidiary role for selecting among technical alternatives of individual components of technological equipment. Under assumptions (a)–(d) (and with the additional condition: [e] constancy of the level of production over the duration of the period of operation of the investment project) the formula for the "absolute" effect (4) becomes:

$$Э = \frac{Ц - C}{E^-} - К. \tag{8}$$

Formulas of type (7) and (8) have been widely extended in the practice of economic calculations, and the "mass consumer" of the recommendations of the "Standard Methodology" scarcely realizes that they are justified only under assumptions (a)–(d), which are usually not fulfilled in the actual economy. Both long-term investments and expenditures on the creation and operation of relatively short-lived elements of fixed capital are estimated as discounted expenditures, and the formula is applied without any modification whatsoever, both for objects with a long construction period and for investments (such as, for example, the purchase of mass-production equipment) where expenditures are almost instantaneously embodied in a facility which is ready for operation. Naturally, the question arises about the degree of error which the application of formulas like (5) and (8) leads to when the actual conditions of the project do not conform to assumptions (a)–(d).

First of all, most facilities being designed are characterized by output variability over the period of operation and a corresponding variability in operating costs. Considering the dynamics of these values, (8) takes on the somewhat more complicated form:

$$\Im = \sum_{t=0}^{\infty} (\text{II}_t - \text{C}_t)(1 + \text{E})^{-t} - \text{K}. \tag{9}$$

This approach to consideration of the future dynamics of the basic indicators (especially operating expenditures in the evaluation of "relative" effectiveness) is used in practice, for example, in railroad project-making. The same method is used to calculate the capital expenditures when they assume a periodic rather than a one-time character. The formula for discounting expenditures, which has been developed to take account of variability of operating costs and periodic capital costs, has the form:

$$- \Im_{\text{cp}} = \sum_{t=0}^{T} \text{C}_t (1 + \text{E})^{-t} + \sum_{t=0}^{T} \text{K}_t (1 + \text{E})^{-t}, \tag{10}$$

and is well known in industries with a relatively high level of skill in the technical-economic appraisal of projects. In principle, the calculating scheme of (10) satisfies the requirements of the theory if we do not consider the vagueness connected with the appraisal of the "calculating period" T and the absence of any kind of instructions for the calculation of depreciation under a finite life.

The second significant simplification of formulas (5) and (8) is the assumption of an infinite operating period for the object. This is quite acceptable if the period of operation is twenty-five to thirty years or more: differences in the values of operating costs calculated by (6) and (7) for such long periods are negligibly small. However, for operating lives which do not exceed ten years, for instance, we cannot ignore these differences.

If the service life of the object is finite and the remaining simplified conditions (b)–(e) are met, the effect of a capital investment is calculated by the formula:

$$\Theta = \text{Ц} - (\text{C} + \text{A} + \text{EК}), \tag{11}$$

where C is the annual operating cost, not including renovation allowances (the latter, naturally, are equal to zero under the infinite project life assumed in (8)), and A is the amortization component of operating costs calculated by the formula:

$$A = \frac{\text{EК}}{(1 + E)^T - 1} \tag{12}$$

Formula (11) emerges from (8) in the following way. From (8), under assumptions (b)–(e) and finiteness of T, we have:

$$\Theta = (\text{Ц} - \text{C}) \sum_{t=1}^{T} (1 + E)^{-t} - \text{К}. \tag{13}$$

We introduce the notation:

$$y = \sum_{t=1}^{T} (1 + E)^{-t} = \frac{(1 + E)^T - 1}{E(1 + E)^T}.$$

Then (14) may be written in the form:

$$\Theta = (\text{Ц} - \text{C})\, y - \text{К}; \quad \frac{\Theta}{y} = (\text{Ц} - \text{C}) - \frac{\text{К}}{y} = \text{Ц} - \text{C} - \text{EК} -$$

$$\text{EК}\left(\frac{1}{E_y} - 1\right) = \text{Ц} - \text{C} - \text{EК} - \frac{\text{EК}}{(1 + E)^T - 1}.$$

The last chain in the equations provides the connection between formulas (7) and (8).

Of the simplified conditions (a)–(e), the most artificial is (c), which assumes that capital investments are embodied in facilities that are ready for operation within an interval of one year. Application of (5) and (8) to real projects characterized by sometimes extremely prolonged construction periods leads to particularly large distortions in the value of the effectiveness of the project or expenditures on its realization.

We will retain conditions (a), (b), (d), and (e), and we will replace condition (c) with the following: the construction period is equal to λ, with a constant amount of capital investment being "assimilated" in each year. Then the very simple formula (8) is written:

$$\Theta = (\text{Ц} - \text{C}) \sum_{t=\lambda}^{\infty} (1 + \text{E})^{-t} - \text{К} \sum_{t=0}^{\lambda-1} (1 + \text{E})^{-t}.$$

We will multiply the right-hand side of this equation by $(1 + \text{E})^{\lambda-1}$. After obvious transformations, we have:

$$\frac{\text{Ц} - \text{C}}{\text{E}} - \left(\frac{(1 + \text{E})^{\lambda} - 1}{\text{E}} \right) \frac{\text{К}}{\lambda}$$

or, finally,

$$\Theta = \text{Ц} - \left(\text{C} + \frac{(1 + \text{E})^{\lambda} - 1}{\lambda} \text{К} \right). \tag{14}$$

This formula for the value of the effect of capital investment undertaken over λ years will have the same structure as (5) and (8). The discounted costs are determined by adding the current and capital expenditures multiplied by some coefficients. A complication which arises in considering the length of the construction period consists in the fact that capital expenditures are discounted not according to the norm of effectiveness, but rather through the multiplier $\dfrac{(1 + \text{E})^{\lambda} - 1}{\lambda}$, the value of which is determined both by the value of E and the construction period λ.

It is useful to introduce a quantitative measure of the effectiveness of the specific investment — the coefficient Θ', determined as the greatest positive x which satisfies the equation:

$$\sum_{t=0}^{T} (\text{Ц}_t - \text{К}_t - \text{C}_t)(1 + x)^{-t} = 0.$$

It is obvious that, for efficient investments, $\Theta' \geqslant \text{E}$, and for inefficient investments, $\Theta' < \text{E}$. In other words, the criteria $\Theta \gtrless 0$ and $\Theta' \gtrless \text{E}$ are equivalent.

The "absolute" effectiveness of capital investment depends primarily on the output prices assumed in the calculation.

125

Assumptions concerning the future facility's product prices are not always sufficiently justified. When comparing production alternatives for a single product it is desirable to include the influence of the price, i.e., to estimate the "relative" effectiveness by the discounted cost per unit of output. If it is necessary to estimate the "absolute" effectiveness of the investment and we cannot be certain that the product price is sufficiently well grounded, we must calculate the effect of its application according to methods which are described below for the production of new products. We will observe that the price review which was conducted in 1967 and later work on the improvement of price formation, especially the forecasting of price dynamics which is being planned, facilitate the problem of correct calculation of the effectiveness of capital expenditures.

4. Special Cases in the Calculation of the "Absolute" Effect

We will return now to the previously mentioned complicated conditions (the facility being designed is intended for the production of a new product which does not have a price; it is necessary not only to evaluate the desirability of the investment, but also to determine its most profitable volume, i.e., to find the optimal intensity of the new production profit; the prices which are used in the calculation depart very substantially from optimal prices, and so on). Unthinking application of stereotyped calculating schemes, when there are notorious and significant differences between the situation under analysis and the assumptions which justify the application of some formula, can only lead to harm, since any solution can only have the appearance of a carefully calculated justification.

We will consider the frequently encountered case in which the result of the capital investment is to create a new product (in the broad sense of the word). Representation of the capital investment as a process of "construction-utilization" of the facility being designed turns out to be impossible, since the

126

components of production cannot be evaluated owing to the absence of prices for them at the time of calculation. In such cases we usually turn to a reduced form of the estimation of the effect, to methods of "relative" effectiveness, leaving aside completely consideration of the problem of the desirability of the capital investment as a source of benefit and limiting ourselves to recommendations of the alternative which will produce the given benefit at least cost. Naturally, there is no guarantee that any of the alternative capital investments is desirable in the first place.

It seems to us that in justifying major investments which have substantial national economic consequences, and, especially, those which lead to the production and utilization of new products, it is especially necessary to demonstrate the full or absolute effectiveness of the project. Calculating the inner product πx in these cases is to use the method that was mentioned above of combining the process $\tilde{x}$, which describes the capital investments (as the totality of construction costs and the differences between the value of the output and operating costs), with the process $\tilde{x}'$, which describes the utilization of new products. In combining the processes $(\tilde{x} + \tilde{x}')$, the new product becomes an internal "intermediate" product with zero external production. In a broader representation such as this, the total effect of production and consumption, rather than the output of a given industry or subindustry, appears as the result of the capital investment.

This approach of combining the construction-production process with a process of utilizing the output by consumers is an appropriate method for measuring the effectiveness of the investment whose results do not have a monetary valuation. The effectiveness of building, for example, a highway in some area cannot be determined by depending only upon the design information which describes the construction process itself. It is easy to calculate the discounted construction and operating costs of the road, but these costs cannot be used for comparison since the use of the road does not cost the users anything and the managerial income from operating it is zero. But, if we

calculate the increase in the profit (savings) of all the users of
the road and add this positive component of the "construction-
operating-utilization" process to the discounted costs (taken,
of course, as negative), the sign of the resulting total will represent the national economic desirability of the road construction
project.

The very same calculation of national economic effectiveness is also necessary in cases in which the marginal operating
costs of the capital construction project are less than the average, and, therefore, the managerial income does not cover
the operating costs, as is usually the case in the optimal utilization of railroads, for example.

The previously introduced ideas of direct investment benefit,
supplementary (or indirect) benefit, and total benefit may be
interpreted in the terms of production processes as inner prod-
ucts corresponding to $\tilde{x}\pi$, $\tilde{x}'\pi$, and $(\tilde{x} + \tilde{x}')\pi$, where $\tilde{x}$ is a vector
describing the given investment as was defined in section 3
and $\tilde{x}'$ is a vector describing utilization of the product of a
given investment by consumers. If we calculate the benefit of
creating a new product, at the design stage the first of the
aforementioned products cannot usually be ascertained owing
to the absence of a price, but then representation of the capital
investments in the form of the expanded vector $(\tilde{x} + \tilde{x}')$ over-
comes this difficulty. Also, in cases in which the existing
prices deviate significantly from optimal prices, it is useful
to set up expanded processes with this calculation so that products with unjustified prices enter in the "intermediate" with
zero external output. In calculations of the effect of the expanded process, unjustified prices cannot then play a part.

Another difficult problem consists of the determination of
the optimal size of the investment. From the theory of convex
programming it is known that if some production process
not belonging to the set X is efficient, i.e., is characterized
by the condition $\tilde{x}\pi > 0$, there exists a number $\bar{h}$ which indicates that the target function of the problem will increase
by a certain amount $\varphi(\bar{h})$ if the process $\tilde{x}$ is included in the
plan with intensity $\bar{h}$. If the intensity of application of process

$\tilde{x}$ is the number h over the interval $[0, \bar{h}]$, the increase in the target function which is achieved by including $h\tilde{x}$ in the plan is equal to $\varphi(h) \leqslant \varphi(\bar{h})$. If the process $h\tilde{x}$ is included in the plan where h is greater than $\bar{h}$, the target function may not only not increase, but it may also decrease. The number $\bar{h}$, which applies to the processes that describe the capital investments, we will call the optimal size of the investment.

The problem of optimal investment size is typical, for example, when determining the desirable shipment of new equipment and when evaluating the effectiveness of expenditures devoted to the creation of new kinds of equipment. The estimation of the latter underlies the material incentive system for workers in project-design organizations which create new machines, equipment, and so on. The practice of performing this kind of calculation in the usual way does not agree with the requirements of theory (or even common sense). Frequently, the effectiveness of utilizing a single sample of new equipment, calculated on the basis of the most favorable circumstances of its application, is multiplied by the total proposed volume, although the effectiveness of utilizing different units of the new equipment may be far from identical (see, for example, [17]).

The principle of determining the optimal volume of shipments is simple, although in many cases it is difficult to apply in practice. It is necessary to study all the possible aids and applications of the new kind of equipment and to calculate the savings which its application provides in all these areas. The annual savings is compared with the value EK, where K is the value of a unit of the new equipment (one of the complications is that this value depends upon the degree of mass production, that is, on the desirable level of the activity). If the savings (the differences between the reduction in annual costs which is brought about by the utilization of the new equipment and the operating costs which are equal to EK) are arranged in decreasing order, the different areas of possible application of the new kind of equipment form a number of groups of application — from those with the greatest net saving to those in which the net saving is negative. The efficient boundary, obviously, is zero. The total demands in all the areas of application with a nonnegative net saving give the optimal size of the investment to be introduced.

For actual calculations of the effectiveness of expenditures in project-design work, the full effect of a new kind of equipment must, it seems to us, be determined by multiplying the norm saving, which is equal to thirty to thirty-five percent (in any case it is less than one half) of the saving from using the machine in the most favorable circumstances, by the optimal volume of its production (the optimal volume of introduction of the investment). The existing practice of multiplying the volume of the investments by the greatest net saving greatly exceeds the effect of project-design work.

A stricter determination of the optimal volume of the investment, of the price of the new equipment, and of the indirect and full benefits involved in its creation may be described within the terminology of convex programming in the following way.

Suppose a model of the economic system is described by a convex programming problem with production possibilities x and optimal prices π. Assume that a new kind of equipment or technology is created which reduces the production costs in various industries and sectors of the economy. The processes of producing and using it, naturally, are not in the set x. They are described respectively by the vectors $(x(h), h)$ and $(y(h), -h)$, where h is an arbitrary positive number which describes the volume (in pieces, for example) of output (utilization) of the new equipment; $x(h)$ is an n-dimensional vector which describes the production process for h units of this equipment; and $y(h)$ is an n-dimensional vector for the utilization of these h units. After considering jointly the process of creating and utilizing the new product, we determine the optimal volume of production and utilization from the equation:

$$(y(\overline{h}) + x(\overline{h}))\,\pi = \max_{h \geqslant 0}\,(y(h) + x(h))\,\pi.$$

In actual conditions the desired maximum obviously always exists, since the costs per unit fall with increasing output at a declining rate and the savings from using the new technology fall inexorably to zero. The (optimal) price of the new technology is equal to $\pi_{n+1} = -x(h)\pi / h$, and the total benefit from creating and introducing it is $(y(\bar{h}) + x(\bar{h}))\pi$, or, what is the same thing, $y(\bar{h})\pi - \pi_{n+1}\bar{h}$.

We should say that, owing to considerations of economic self-sufficiency, for a single unit of the new equipment or technology it is more correct to establish a price at the level of the effect of the marginal unit $y'(h)\pi = -x'(h)\pi$, since, with such a price, its acquisition will be justified by the benefit of applying it by each consumer (which will also be beneficial from the point of view of the national economy). Production costs will be only partially covered; and for the rest, the effectiveness of the expenditures on design, execution of the product, and production of new equipment will be determined by the national economic consequences of its use.

When the undertaking of new production is a prolonged process during which production costs fall, the area of application and the number of users of the new product will expand and the optimal rate of development of the new production must be determined. It must be determined by the principle of the greatest benefit from use of this volume of output, less costs, summed over the entire period, and discounted to the initial period. As a rule, the optimal solution cannot be completely realized by cost-accounting, and a reduced price for new products for stipulated periods must be justified by the national economic benefit for its users. Only in orienting ourselves by the total, national economic benefit, will the new product be produced at rates corresponding to the task of total realization of the possibilities of technological progress. Estimation of the effectiveness of this kind of undertaking must be oriented not to financial indicators of enterprises which produce the new equipment, but must be conducted on a wide scale, with consideration of the influence on the whole national economy over a long time-span.

The methodological principles of estimating the effectiveness of capital investment which have been presented in this paper may be used not only to analyze capital investment projects proper, but may also be used to estimate the effectiveness and optimization in conducting any kind of long-range program in which the time factor must be taken into account. This applies, for example, to the creation of reserve stocks and reserve capacities, the extension of new kinds of consumption products and services, new scientific-technical developments and their introduction into production, measures for preservation of nature and rational schemes for the exploitation of natural resources, rates and sequence of industrial assimilation of new regions, and so on.

In connection with the problem of the optimal size of a capital investment, we note that most construction projects uniquely define the scale of output of the future production of the object. The goal of the economic analysis of projects in such cases is not the determination of the optimal volume of the investment project but analysis of the desirability of undertaking the project with the scale of output which is being envisaged. However, the problems connected with the size of the investment projects indirectly arise in these calculations as well.

The calculation of $x\pi$ for process x, which describes the directly given capital investment, or for the component-expanded vectors which include the indirect and supplementary benefits, depends upon the assumption of stability of the system of prices π and the suitability of this system for the description of a situation which will arise after execution of the planned capital investment. Meanwhile, of course, it is possible to have capital construction projects which will lead to a radical change in the balance of production and consumption of various products and, consequently, in the system of optimal prices. Strictly speaking, the methods proposed in this paper for determining the effectiveness of capital investment, for example, the simplified methods which are recommended by existing instruction documents, are applicable only for "small" investments whose undertaking does not lead to radical changes in the plan and the system of prices which characterizes it. For "large" investments, naturally, it is not possible to indicate approaches for calculation which rely on local information to describe the national economic consequences of the adoption of the project. "Large" investments by definition cannot be described by any local information different from the information which describes the plan as a whole; therefore, for "large" investments the problem of effectiveness, as defined in section 1 above, does not exist. Their desirability and effectiveness may be determined only by optimization of the plan as a whole — comparing plans including and not including the measure in question.

Notes

1) It is notable that an immeasurably greater literature is devoted to the norm of effectiveness of capital investment than to the problem of measuring the economic value of any other resource, including such universal resources as labor, land, water, etc.

2) If the prices π are not uniquely determined, it is possible to have cases in which it is impossible to improve the solution by means of $\bar{z}$ even though $\pi \bar{x} > 0$.

References

1. L. Iushkov, "The Basic Issue in Planning Methodology," Vestnik finansov. 1928, no. 10.

2. S. A. Kukel'-Kraevskii, "A Generalized Method for Selecting the Optimal Parameters in Electric Installations," Elektrichestvo, 1940, no. 8.

3. M. M. Protod'iakonov, The Study and Design of Railroads, Moscow, Tranzheldorizdat, 1934.

4. A. L. Lur'e, "On the Economic Evaluation of Technological Measures," Tekhnika zheleznikh dorog, 1946, no. 5-6.

5. A. L. Lur'e, "Methods of Comparing Operating Costs and Capital Investments in the Economic Evaluation of Technological Measures," in Problems of Railroad Economics, Moscow, Tranzheldorizdat, 1948.

6. A. L. Lur'e, On Mathematical Methods of Solving Optimum Problems in Planning the Socialist Economy, Moscow, "Nauka" Publishers, 1964.

7. V. V. Novozhilov, Problems of Measuring Costs and Benefits in Optimal Planning, Moscow, "Nauka" Publishers, 1964.

8. L. V. Kantorovich, Economic Calculation of the Best Use of Economic Resources, Moscow, USSR Academy of Sciences Press, 1959.

9. V. N. Bogachev, "The Recoupment Period": The Theory of Comparing Planning Alternatives, Moscow, "Ekonomika" Publishers, 1966.

10. V. N. Bogachev and L. V. Kantorovich, "The Price of Time," Kommunist, 1969, no. 9.

11. "Standard Methodology for Determining the Economic Effectiveness of Capital Investments in New Technology in the National Economy of the USSR," Moscow, Gosplanizdat, 1960.

12. "Standard Methodology for Determining the Economic Effectiveness of Capital Investments in New Technology in the National Economy of the USSR," Ekonomicheskaia gazeta, 1969, no. 39.

13. L. V. Kantorovich, "On a Method of Solving Certain Classes of Extremal Problems," Papers of the USSR Academy of Sciences, 1940, no. 28.

14. L. V. Kantorovich, "A Dynamic Model of Optimal Planning," in Planning and Mathematical-Economic Methods, Moscow, "Nauka" Publishers, 1964.

15. L. V. Kantorovich and V. L. Makarov, "Optimal Models of Long-Range Planning," in Applications of Mathematics in Economic Research, vol. 3, Moscow, "Mysl'" Publishers, 1965.

16. A. L. Lur'e, "Optimal Prices and the Norm of Effectiveness," Ekonomika i matematicheskie metody, 1967, vol. III, no. 2.

17. L. V. Kantorovich and V. A. Bulavskii, "Possibilities for Utilizing Mathematical Optimal Planning in the Analysis of Effectiveness of New Technology," in Materials of the VISKhOM NTS, no. 18, Moscow, 1964.

11

A One-product Dynamic Model of the Economy Taking into Account the Change in the Structure of Capital under Technological Progress*

L. V. KANTOROVICH and V. I. ZHIIANOV

We consider an economic system in which one product is produced a part of which is used for consumption and a part of which is used to increase the basic capital stock and circulating capital [1].

Let $T(t)$ be the labor supply at time t and $K(t)$ be the basic capital (nominal) at time t. We introduce the parameter $\lambda(t)$ for the type (or structure) of new capital that describes the price (expressed in physical products) of normalized capital (capital per unit of labor), or in other words, the degree of capital intensity of labor. We assume that capital created at time t is of the same type ($\lambda(t)$ is a single-valued function of t).

By $\varphi(t)$ we will denote the intensity of new capital creation, i.e., $\varphi(t)\,dt$ is the number of new work slots created during time $[t,\ t+dt]$, so that $\lambda(t)\varphi(t)\,dt$ is the volume of newly created capital during the same interval. The functions $\lambda(t)$ and $\varphi(t)$ (and $K(t)$) must be determined in the model.

We will assume that the possible production processes are described by a production function $U(x, y)$ that shows the quantity of net output created by labor y by using basic capital x per unit of time (at the starting point). It is assumed that the

---

*Doklady Akademii nauk SSSR, 1973, vol. 211, no. 6.

function $U(x, y)$ is a positive, homogeneous, first degree function

$$U(\lambda x, \lambda y) = \lambda U(x, y) \text{ for } \lambda > 0,$$

and is based on optimal processes, which makes it natural to assume convexity of $U(x, 1)$.

Technological progress can be taken into account in the model in the following way. It is assumed that the quantity of net output produced per unit of time with a given capital stock and labor input grows exponentially as a function of the time of creation of capital stock τ and is $e^{\delta\tau}$ times as great as the output produced by means of the capital created at the initial moment of time under the same conditions.

It is also assumed that in the course of development of the economy there is a withdrawal of labor from capital having a low structure — what was created long ago. The released labor resources are used with the newly created capital. The discarded capital is not used in the future. As will be evident from what follows, this assumes that for the function $\varphi(t)$ the condition $\varphi(t) \geqslant T'(t)$ for $t \geqslant t_0$ holds.

Suppose that now the set E_t, which belongs to the interval $[0, t]$, is defined such that $\tau \in E_t$, if capital created at time τ, $\tau < t$ is retained and used up to time t.

Then the quantity of net output (national income) that may be produced at time t per unit of time is

$$P(t) = \int\limits_{E_t} e^{\delta\tau} U[\lambda(\tau)\varphi(\tau), \varphi(\tau)] d\tau. \tag{1}$$

In assuming continuous growth of the organic composition of capital, the structure of the set E_t is as follows: suppose that by time t all capital which has been created before some given moment of time $m(t)$ has been released; then E_t represents the interval $[m(t), t]$. The function $m(t)$ must be determined in the model.

Capital investment used to increase basic and circulating

capital is given by its intensity, so that $\varkappa(t)\,dt$ is the volume of capital investment over the time interval $[t,\ t + dt]$. The functions $\varkappa(t)$ and $T(t)$ are given in the model, and $\varkappa(t)$ may be made to depend on $P(t)$ or other parameters in the model.

We will write the equations of the model.

The labor balance. The incremental demand for labor $\varphi(t)\,dt$ at time t is satisfied partly through the increase in labor supply and partly by means of the labor released from the capital that has been written off. We have the following equation:

$$\varphi(t)\,dt = T'(t)\,dt + \varphi[m(t)\,]m'(t)\,dt. \tag{2}$$

The capital balance. The volume of newly created capital during time dt is equal to $\lambda(t)\varphi(t)\,dt$. It is created by capital investment, so that

$$\lambda(t)\varphi(t) = \varkappa(t). \tag{3}$$

The differential optimization condition. This condition means that an increase in net output at each moment of time must be maximal (maximization of net output during some planning period $[t_0,\ t_1]$ or some other global optimization condition may be another possible optimality criterion). In other words, the functions $\lambda(t)$, $\varphi(t)$, and $m(t)$ must be determined in such a way that $dP(t)\,/\,dt$ is maximal at each moment of time t. Considering $dP(t)\,/\,dt$ as a function of the aforementioned variables, and considering (2) and (3), we get the differential optimization condition in the form

$$\varphi(t)\,U\,[\lambda(t),\ 1] - \varkappa(t)\,U'_x\,[\lambda(t),\ 1]$$
$$- e^{\delta\,[m\,(t)-t]}\varphi(t)\,U\,[\lambda\,[m\,(t)],1] = 0. \tag{4}$$

Thus the system of equations that describe the model consists of (2), (3), and (4). The system is solved for $t > t_0$ (t_0 is a fixed number). The initial conditions are given in the form

$$m(t_0) = m_0, \quad \varphi(t) = \varphi_0(t), \quad \lambda(t) = \lambda_0(t) \text{ for } t \in [m_0,\ t_0],$$

137

where m_0 is a given number, and $\varphi_0(t)$ and $\lambda_0(t)$ are functions that give the initial distribution (with $t \in [m_0, t_0]$) of capital and labor.

The case of no foreign capital investment. Suppose that $V(t)$ is the part of net output used for consumption, and the other part is used to create basic capital. Equation (3) takes the form

$$\lambda(t)\varphi(t) = P(t) - V(t).$$

Usually consumption is assumed equal to a fixed share of net output

$$V(t) = (1 - \gamma)P(t)$$

($0 < \gamma < 1$ is a given number).

Calculation of the period of capital creation — lags. We will now assume that the capital investment is not turned instantaneously into capital stock, i.e., we take into account the length of the construction period and the assimilation of capital. We will assume a period of construction and assimilation of capital equal to L. Obsolescence of capital during the period of construction and assimilation is considered in the following way: capital which has gone into operation at time t has the structure $\lambda(t)$ and a level of technological progress represented by the multiplier $e^{\delta(t-L)}$. For the same reason, the volume of capital which has gone into production (during the interval $[t, t + dt]$) is determined by the intensity of capital investment at time $t - L$ and is equal to $\varkappa(t - L)dt$.

The equations of the model in this case take the form

$$P(t) = \int_{m(t)}^{t} e^{\delta(\tau-L)} U[\lambda(\tau),1]\, \varphi(\tau)d\tau, \qquad (1a)$$

$$\varphi(t) = T'(t) + \varphi[m(t)]m'(t), \qquad (2a)$$

$$\lambda(t)\varphi(t) = \varkappa(t - L), \qquad (3a)$$

$$\varphi(t) U[\lambda(t), 1] - \varkappa(t - L) U_x'[\lambda(t), 1]$$
$$- e^{\delta[m(t)-t]} \varphi(t) U[\lambda[m(t)], 1] = 0. \qquad (4a)$$

138

A Dynamic Model and Technical Progress

The norm of effectiveness of capital investment (in the future we will denote this parameter by $n(t)$) is determined by the growth (per unit of time) of produced output divided by the incremental capital investment which has supported it [2].

In this model we will consider the following economic maneuver. Suppose that during the course of a very short interval of time dt we get incremental means of production in amount dx which must be returned during some period of time Δt. We will use the incremental capital to create during time dt a capital stock of higher structure ($\lambda(t + \Delta t)$ instead of the planned structure $\lambda(t)$), which will give the accumulation of the incremental product. Furthermore, during the period $[t + \Delta t, t + \Delta t + dt]$ capital stock with a structure $\lambda(t)$ is created that permits us to return the additional investment earlier. As a result of this process, by time $t + \Delta t + dt$ we will have a spectrum of the structure of the capital stock and a volume of capital investment that practically coincides with what is generated during the development of the economy without additional capital investment. Calculating the efficient growth of net output in relation to the incremental capital investment (per unit of time), we get the formula

$$n(t) = e^{\delta t} U'_x [\lambda(t), 1],$$

to calculate the norm of effectiveness, or, using equation (4),

$$n(t) = \frac{1}{\varkappa(t)} \left\{ \frac{dP(t)}{dt} - e^{\delta m(t)} U[\lambda[m(t)], 1] T'(t) \right\}. \quad (5)$$

In other words, $n(t)$ is determined by the ratio of the growth of national income, from which we deduct the volume of possible net income from the new entrants to the labor force when working with written down capital (the structure $m(t)$), to the value of the capital that has come in during the time interval.

Using (5) we get the relationship $dP / dt = \varkappa(t) n(t) + T'(t) \partial_T$, from which it is obvious that the growth of national income is determined by the sum of newly introduced capital, multiplied by its effectiveness, and the incremental labor, multiplied by its marginal effectiveness ϑ_r (the effectiveness with the released capital).

Solution of the equations of the model by the method of small parameters. At this point we will give a concrete form to the functions $U(x, y)$, $T(t)$, and $\varkappa(t)$. The production function has the form $U(x, y) = x^\alpha y^{1+\alpha}$— a Cobb-Douglas function. Labor reserves grow exponentially $T(t) = e^{pt}$, p is the rate of population growth. The density of capital investment will be assumed to be constant, $\varkappa(t) = \varkappa$, where $\varkappa$ is a given positive number.

We will assume that the size of p and δ (δ is an index of technological progress) is small, which agrees with statistical data. We will assume that the system of equations (2)–(4) has a solution, and the functions $\lambda(t)$, $\varphi(t)$, and $m(t)$ may be expanded in power series of p and δ:

$$\lambda(t) = \bar{\lambda}(t) + p\lambda_p(t) + \delta\lambda_\delta(t) + p^2\lambda_{p^2}(t) + p\delta\lambda_{p\delta}(t) + \delta^2\lambda_{\delta^2}(t) + \ldots,$$
$$\varphi(t) = \bar{\varphi}(t) + p\varphi_p(t) + \delta\varphi_\delta(t) + p^2\varphi_{p^2}(t) + p\delta\varphi_{p\delta}(t) + \delta^2\varphi_{\delta^2}(t) + \ldots,$$
$$m(t) = \bar{m}(t) + pm_p(t) + \delta m_\delta(t) + p^2 m_{p^2}(t) + p\delta m_{p\delta}(t) + \delta^2 m_{\delta^2}(t) + \ldots$$

We put these series into the system (2)–(4) and equate terms with identical order of smalls (we neglect terms of higher than the first order of smalls relative to p and δ), and solving the resulting equations, we determine the function $\bar{\lambda}(t)$, $\bar{\varphi}(t)$, $\bar{m}(t)$, $\lambda_p(t)$, $\varphi_p(t)$, $m_p(t)$, $\lambda_\delta(t)$, $\varphi_\delta(t)$, $m_\delta(t)$.

The solution of system (2)–(4) under the assumptions made at this point takes the form

$$\lambda(t) = \frac{\varkappa t}{\beta}\left\{1 - \delta\frac{t - t_0}{2a} - p\frac{t - t_0}{\beta(1 - \beta)}\right\},$$
$$\varphi(t) = \frac{\beta}{t}\left\{1 + \delta\frac{t - t_0}{2a} + p\frac{t - t_0}{\beta(1 - \beta)}\right\},$$
$$m(t) = \beta t\left\{1 + \delta\frac{(1 - \beta)t}{2a}\right\},$$

where $\beta = (1 - \alpha)^{1/\alpha}$.

Using this solution, we get for the norm of effectiveness of capital investment the following expression:

$$n(t) = Mt^{\alpha-1}\{1 + \delta[at^\alpha + b(t - t_0)] + p[c(t - t_0)]\},$$

A Dynamic Model and Technical Progress

where

$$M = \frac{a\varkappa^{1-\alpha}}{\beta^{1-\alpha}}, \quad a = \frac{\varkappa^{1-\alpha}}{\beta^{1-\alpha}}, \quad b = \frac{1-\alpha}{2a}, \quad c = \frac{1-\alpha}{\beta(1-\beta)}.$$

References

1. L. V. Kantorovich and L. I. Gor'kov, DAN, 1959, vol. 129, no. 4.
2. L. V. Kantorovich and Al'b. L. Vainshtein, Ekonomika i matem. metody, 1967, no. 5.

12

Optimal Mathematical Models in Planning the Development of a Branch and in Technical Policy*

L. V. KANTOROVICH

The return to the branch management of industry and its elevation to a new level create conditions favorable to the scientifically substantiated resolution of the problem of branch development as a whole in all its aspects. In this area considerable use must be made of mathematical methods and models as well as of analysis and calculations based on them. At the present time rather widespread use is being made of the simplest optimal models of the development and location of branches as well as models calculated by linear programming methods and in many cases resolved even on the basis of algorithms and the transport problem. (1) Calculations of the development of branches as a whole or in individual regions have already been made for a number of branches with the aid of these models, and in preparing the plan for the next five years they must encompass a greater portion of the branches. The results of calculations for certain branches (e.g., the fuel-energy balance) were already considered in the compilation of the five-year plan for 1966-70.

The basic advantage of these models over the traditional system of isolated economic analysis for individual enterprises

---

*Voprosy ekonomiki, 1967, no. 10.

and of the balancing of production and demand is the complex approach in which there is a simultaneous joint analysis of the nation's total demand and of the production potentialities of all enterprises in a branch — existing and projected. This makes it possible to achieve an optimal combination of the interests of producers and consumers, the maximum national economic effect, and to obtain new, valuable economic indices and characteristics.

In this way mathematical models make possible the fullest realization of the advantages of a planned socialist economy and centralized planning in the determination of basic directions in the development of a branch. On the other hand, we should like to point out that the application of optimal mathematical methods is by no means confined to models of long-range planning of branch development. They can be equally applied in current and day-to-day planning, in the solution of problems in technical and economic policy, management, etc. Moreover, we should not confine ourselves to standard models and methods or exaggerate their significance. Various branches — extractive and manufacturing, with large and small product lists, with complex and simple technological relationships, with continuous mass and series production, etc. — have to a very great extent their own specific features. A model of a branch will be sound and effective only if it is built on the basis of a thorough understanding of production processes, of the structure and interrelations of the branch, if it is worked out by specialists who take into account the particular features of the technology and economics of the branch to the same extent that they consider the possibilities of mathematical model-building. The purpose of standard mathematical models and methods is to provide orientation with respect to the direction and nature of this research, to provide certain instructions and specimens, to pose and outline the solution of general methodological questions. The newness and imperfection of these models notwithstanding, the advantage they give is unquestionable and, to all appearances, completely makes up for those additional possibilities that the concrete economic analysis of individual enterprises gives. Nonetheless, we must also see the shortcomings in these models. A considerable part of the present article is devoted to the analysis of some of these models, to posing the problems, and to methods for improving these models.

Moreover, it seems advisable that the modeling of the development of a branch, like other problems of the application of mathematical methods, be carried out in each branch by a collective of branch specialists systematically engaged in work on this complex of problems. Use must also be made of the scientific influence of scientific institutions dealing with general methodological problems. This influence should be in the form of methodological and advisory assistance, in joint elaboration, or in some other possible form. Mathematical optimal models must be skillfully combined with traditional methods of technical-economic analysis and planning.

A Model of Development of a Branch and Possibilities for Improving This Model

The basic, most widely used, and simplest model of the development of a branch is calculated for a given period in the future; in the simplest instance the production of a single commodity is involved. The future demand for a commodity is assumed to be given and localized. The locations of existing enterprises, the possibility of expanding them, and the admissible locations of new enterprises are indicated. In all cases possible dimensions of commodity output and the enterprise cost of production are established for existing capacities, while for new capacities calculated outlays and permissible volumes are given. (By calculated outlays we usually mean the sum of $\sigma = s + \gamma k$, where s is the enterprise cost of production, k is capital investments per unit of output and γ is the effectiveness norm.) The problem is to minimize total outlays for production, outlays for capital investments (calculated in accordance with the effectiveness norm), and outlays on the transportation necessary to satisfy the demand for the commodity. In the final

144

Optimal Models in Planning

analysis it is written in the form of a linear programming problem of a special type, which reduces to the transport problem. Its mathematical formulation is as follows. Find

$$\min_{x_{i,k};\ y_i;\ z_i} \sum_{i,k} c_{i,k} x_{i,k} + \sum_i s_i y_i + \sum_i \sigma_i z_i$$

with the conditions:

$$1)\ \sum_i x_{i,k} \geqslant B_k;\ 2)\ 0 \leqslant y_i \leqslant L_i;\ 3)\ 0 \leqslant z_i \leqslant M_i;$$

$$4)\ \sum_k x_{i,k} = y_i + z_i;\ 5)\ x_{i,k} \geqslant 0,$$

where s_i is the enterprise cost of production at an existing enterprise at point i; σ_i is the calculated expenditures at a new enterprise in point i; L_i and M_i are the permissible production capacities of enterprises; y_i and z_i are the planned production volumes; B_k is the demand for a given commodity at point k; $c_{i,k}$ are expenditures on transporting a unit of output from point i to point k; and $x_{i,k}$ is the planned volume of shipments from point i to point k.

The above requirements indicate respectively: (1) the satisfaction of demand, (2) and (3) restrictions on production capacities, (4) the balancing of transport flows and output volumes, and (5) the nonnegative nature of transport flows. Such a model can be successfully calculated on computers even if there are several hundreds or thousands of points. (2)

In the case of the production of several types of commodities and the possibility of their joint production, as well as in the case of other types of restrictions and other optimality criteria, the problem no longer reduces to a transport problem but to a more or less general linear programming problem (if the hypotheses of linearity are still observed).

Despite the above-mentioned great advantages of compiling the plan on the basis of a single model as compared with the traditional method of isolated economic analysis for individual

145

enterprises, the utilization of such models encountered considerable difficulties even for those branches which are simplest in terms of structure (e.g., cement production). These difficulties are related not only to the conservatism of a certain element of the personnel and to the unusual nature of these methods, or to the absence or poor quality of the necessary data, but also to certain shortcomings and imperfections in the models themselves and in the methods for calculating and analyzing them. Accordingly, it is important to analyze these shortcomings and difficulties and to indicate ways of improving the models and the indices they use; it is necessary to outline the mathematical, economic, and statistical research and means required to bring them closer to reality.

The transition to dynamic models. The models under investigation are semidynamic in nature since they are only calculated for a given time span and, accordingly, do not take into account the necessary growth of rates of production and limitations with regard to means. Nor do these models take into account the economics of the location of production facilities during the period. For a number of years now, at the Institute of Mathematics of the Siberian Department of the Academy of Sciences of the USSR, a proposal has been advanced concerning the widespread utilization of optimal dynamic models, balanced for each year in the planned period, and particularly applicable to branch models. (3) Effective methods for calculating such models have also been devised. For example, for the simplest cases an algorithm similar to the transport one has been proposed by V. L. Makarov and also by S. S. Surin (Leningrad State University). Some experimental calculations have been carried out.

The use of the dynamic model can produce a more effective plan of branch development which makes it possible to avoid the aforementioned shortcomings. Such a dynamic model can encompass not only the planned period but also a period beyond that referred to in the balance [zabalansovyi period]. This, taking into account the forecast of norms of outlays and requirements, makes it possible to calculate a certain forecast with

146

respect to the development of the branch (this may be in the form of variants or it may be stochastic). In this way this model makes it possible to construct a plan that is not only coordinated with the condition of the branch during the present period but also with the forecast of its future development.

It should be pointed out that actual production is frequently more dynamic than presented in this model. For example, the hypothesis adopted in the basic model as to the constancy of production capacities and outlays at existing enterprises frequently does not correspond to reality. As experience shows, under the conditions of a complex production structure, the volume of production generally increases with time while expenditures decline, with relatively small outlays being made for modernization and technical progress. Such patterns should be studied concretely branch by branch and they should be considered in the model on a single-valued or variant basis. Dynamic production functions of the branches as a whole and for individual production facilities must be plotted and used.

Finally, it is not advisable to insert the complete description of certain production facilities with a complex structure in the mathematical model. Rather, it is preferable to combine this approach with the methods of variant technical-economic planning. For example, proceeding from several traditional variants (labor-intensive, capital-intensive, forced, slowed-down, etc.) of drafts of the overall development of a number of enterprises, combines, or other organizations, the optimal combination for a branch as a whole should be sought on the basis of constructing a mathematical dynamic model. Some simplified calculations of this type have been done at the Laboratory of the Economic-Mathematical Institute [LEMI] of the Siberian Department of the Academy of Sciences of the USSR.

Consideration of nonlinearity and indivisibility. As we know, the constancy of expenditures per unit of output, assumed in the basic model, is not always observed; moreover, the dependence between expenditures and output is frequently nonlinear. In many branches it is also necessary to consider the indivisibility of certain objects or individual aggregates which can only be

147

incorporated in the plan in their entirety. This complicates the calculation of the model and brings it into the area of nonlinear or integer-valued programming. However, in view of the special nature of these problems, in a number of instances they are successfully resolved — either precisely or approximately — with the aid of linear programming algorithms or simple modifications of them. In other cases, considering the specific features of the problems, it is possible to resolve them considerably more simply than general problems of nonlinear and integer-valued programming by using dynamic programming methods, partial scanning, and others, in conjunction with linear programming methods and qualitative economic analysis. Other effective devices are certain approximate and iterative methods, the description of which was given in a number of papers at the All-Union Conference on the Application of Mathematical Methods in Branch Planning and Management (December 1966). Thus we should not exaggerate the mathematical difficulty of these problems; this is also confirmed by the work of the institute research laboratory. (4)

However, considerable difficulties are entailed in the construction of the initial economic nonlinear dependences of outlays on the volumes of production. There are no sufficiently satisfactory methods for establishing these dependences. Some progress in this regard can be expected from work being carried out on improving prices and the related more extensive and complete analysis of outlays.

The incorporation in the model of the dependence of the volume of demand on the size of outlays or on the selling price would not present any substantial difficulties for the construction and calculation of this model. However, in this case it is still more difficult to obtain substantiated dependences. In such problems in which the volume of output itself must also be defined, it is not the enterprise cost of production that must be juxtaposed against the effect for the consumer but rather the total outlays on commodity production, in particular, differential outlays, while in dynamic models it is the dynamics of these outlays and the corresponding shadow prices.

Multiproduct models. In planning the development of a branch whose enterprises produce many types of commodities, the use of optimal models becomes considerably more complicated. The calculation of such models is more difficult and is no longer reduced to the transport problem but instead usually requires general algorithms of linear programming; there is also a corresponding increase in the complexity of taking account of dynamics, nonlinearity, and indivisibility. However, the basic difficulty is that of obtaining the initial information. Primary importance is attached to the successful aggregation of production. Also needed are methods for determining the productive capacity of an enterprise, making it possible to establish the potential for producing a complex of products, norms of expenditures under a given production program, the forecast of commodity structure required by the consumer, etc. In this case the model may also include several alterations and partial interchangeability of commodities with respect to the consumer. Owing to the uncertainty of some of the data (e.g., the structure of demand), the methods of stochastic programming may be used for the solution. In particular, this approach is essential for determining reserves of capacities. A certain amount of experience in making calculations for multiple products has been accumulated in the production of building materials and structural components and in other branches.

Frequently, in place of taking into account the possibility of a continuous change in the composition of the production program, in these models it is necessary to proceed from several discrete variants in the organization of the enterprise's technology. The number of these is extremely limited since each variant requires that an individual technical plan, estimate, etc., be worked out. It seems to us that in the future the automation of technical planning and calculations could be used to this end. If they are algorithmized and fed into a computer, it will not be difficult to obtain data on dozens and hundreds of possible variants.

Resources, raw materials, supplies. In a number of branches, commodity production is so directly connected to the raw material base that it is advisable to include their entire complex in the model. Thus a given type of enterprise and technology must be selected with regard to the possibilities for raw materials. Further, one cannot rely, for example, on established standardized periods for the delivery of supplies (with respect to mines, oil extraction, and ores). Rational periods for fabrication and the sequence in the utilization of raw material resources must be established in the very process of calculating the optimal dynamic model for an extended period of time, taking into account the norm of effectiveness and the calculated relationship between inputs and outputs at different times. As a rule, such calculations select shorter periods and more forced methods of elaboration than those recommended by approved norms and norms that have been established in practice. In a number of cases this also leads to the necessity of establishing greater production capacities in the extractive complexes and to a greater concentration of means in space and in time.

Consideration of local conditions. In selecting the rates of development of production and the size of the enterprises, it is necessary to consider different variants with respect to the type of technology, the dimensions and degree of restructuring and integrating, the volumes of capital investments, and labor intensiveness. In particular, this is important when considering local conditions — the shortage or surplus of manpower, the availability of housing, construction possibilities, and the energy balance. The economic form of considering these differences will become more effective as price formation and other economic evaluations improve; nonetheless, economic evaluations will not always reflect these differences with sufficient completeness and accuracy. In such cases local conditions must be considered in a special way — in the form of one or another restriction, fine, benefit, etc.; but the effort must be made to see to it that local conditions are also reflected in economic indices and in cost accounting. This, for example, applies to a proposal — more than once advanced by us and now receiving broad support — concerning the establishment of payments for the use of labor groups of which there is a shortage and subsidies for using surplus labor groups or in regions with surplus manpower resources. (5)

The requirement that local conditions be reflected in economic indices also applies to other aspects. For example, it is sometimes advisable to maintain an existing enterprise which has obsolescent equipment. However, it will function normally and without losses only if the necessary adjustments are made in cost-accounting, finance, and valuational indices (freeing the enterprise from making payments for funds and from making depreciation payments or a sharp decrease in these; subsidies for utilizing manpower, etc.). It is also necessary to introduce differentiated selling prices on building materials and construction works, depending on their existing balance, etc.; otherwise there may be a gap between the optimal plan decisions and cost-accounting conditions, as a result of which the rationally selected enterprises may prove to be unprofitable. The question of transport rates is particularly acute in this respect.

Transport outlays. In models of branch location, the consideration of transport outlays which play a very substantial role is extremely unsatisfactory. Possible solutions — rates, enterprise cost of production, calculated outlays, differential outlays (dependent outlays and dependent capital investments) — yield figures which differ by several times among themselves. The most optimal are the differential outlays, since the inclusion of one or another location variant, which is connected with an increase in shipping, engenders national economic expenditures on the increase in shipping, i.e., differential outlays. However, in addition to the level of average outlays, the proper analysis of outlays for various types of shipping (distance, type of freight, region) is of great importance. For example, the per ton-kilometer outlays connected with a 700 km increase in the distance of shipping over the average distance of a shipment are almost twice as low as the average outlays for this same distance. This conclusion is confirmed both by the statistical analysis of data on our rail shipping as well as by the relationship of rates according to rail rate schedules of the USA (in

151

constructing the latter the greater elasticity of demand for long-distance shipping is also considered, but this circumstance plays a certain role in our case as well). At the same time, our methods for calculating enterprise cost of production — based on leveling average indices which do not consider actual current and fund outlays on a differentiated basis — do not reveal the correct relationships of outlays for shipping. Thus the transport ministries and research institutions are confronted with the important problem of more deeply investigating the structure of transport outlays, taking into account the economic reform that is being carried out and the principles of optimal planning. These outlays must be expressed in sufficiently generalized and aggregated form, but with the necessary differentiation.

As we know, for various reasons in the process of reviewing prices, major changes are not being made in transport rates. But in the future, changes in the level of rates and the regularization of rates (which will bring them closer to real national economic outlays on shipping [differential] on the basis of extensive study of the structure and size of transport outlays) will be absolutely essential. Otherwise, the inclusion of outlays in optimal models of location without a regularization of rates will lead to a gap between plan decisions and cost-accounting conditions.

Thus the regularization of transport rates is a necessary prerequisite to the rational solution of problems of location for many branches of production. However, this is important not only for long-range but also for day-to-day planning. Incorrect consideration of transport outlays engenders superfluous production outlays, the underutilization of fixed capital, unsubstantiated demands for new capital investments, etc. This is why the proposal of V. Cherniavskii that, in siting production facilities, transport be considered in terms of total calculated outlays is unacceptable. (6) Such an approach does not consider the specific features of the transport branch: the sharp non-linearity in expenditures — of current and, even more, capital expenditures. Accordingly, this approach is not scientifically based and in practice it incorrectly orients the planning and economic organizations toward superfluous capital investments in enterprises that duplicate one another or in enterprises with natural sources of raw materials that are not favorable for development.

Structure of output. Generally, in models of branch development, structure of output and the need and demand for it are regarded as given. At the same time, major problems in branch development are the improvement of this structure, its dynamics, changes in the proportions, and the conversion to types of commodities which are more progressive and effective for the consumer and which consider new requirements and are more advantageous in production. All these things must also be considered in the model, for which purpose, to the extent that objective methods permit, a study should be made of the areas of effective utilization of concrete types of commodities, of their consumer effectiveness and evaluation. In the case of consumer goods it is important to study demand and its elasticity. Data on the conditions of substitutability of commodities are needed. Finally, an objective assessment must be made of changeability in demand per se and the extent of such changes. The structure of production, production capacities, and their universality must be verified against the possibility of an effective consideration of such changeability. This is accomplished, for example, by the proper combination of universal and special-purpose equipment, by the existence of parameters on the management and restructuring of production. Statistical analysis and stochastic models must be used here. At the same time, in this area we can only name individual attempts at research (production of plastics, oil refining).

It seems to us that the dynamics, diversity, and possibility of a flexible operational change in assortment in accordance with changes in demand and the requirements of the consumer are an important qualitative characteristic of the production potential of the branch (probably admitting of objective evaluation as well). It must necessarily be considered together with indices of the growth of production volume.

Thus demand is of decisive significance in determining the branch plan. This once more emphasizes the timeliness of scientific elaboration of consumption problems, which are now attracting a great deal of attention on the part of political economists and mathematical economists.

Coordination of the branch model with other models. In constructing a model of branch development it is necessary to consider two opposing tendencies. On the one hand, the structure of the model must be as autonomous as possible, with minimal and most simplified requirements with respect to the data of other branches, economic regions, and the national economy as a whole, relating to the structure and volume of production, to demand and economic characteristics. Therefore, the latter must be chiefly given in a largely aggregated valuational form. On the other hand, if the model is to be improved, more concrete and detailed information on the contiguous branches, on regional balances, etc., is required. To a considerable extent, the quality and practicability of the model will depend on the successful combination of these requirements, taking into account the specifics of the branch and its relationships. Another means of overcoming this difficulty is the integration of closely interrelated branches (the fuel-energy balance). A radical solution to this problem lies in the implementation of an overall system of optimal planning, which will lead to the simultaneous development and joint examination of a complex of models for a number of branches. However, even before the realization of this stage, it is necessary that the structure of each model provide for the possibility of its being incorporated in a complex, that each model have the necessary free parameters of inputs and outputs which for the time being could be filled out in conditional and simplified form on the basis of preliminary (not optimal) draft plans and forecasts.

Questions for discussion. Since much attention is focused on mathematical models of planning of development of a branch both in theoretical investigations as well as in practice (these investigations and calculations have been applied to various branches by different collectives and organizations), it is natural that there be a divergence of views on the structure of the models and on the methodology of the calculations. In our estimation, the principal objective cause of this discussion is that now the optimal models of branch development are being applied prior to the establishment of an overall system of optimal planning at all levels and in all links of the national economy in the face of the absence of a number of economic indices connected with the overall optimal plan (with other branches, etc.). If the system of optimal planning were introduced everywhere, the structure of the branch model would be essentially indisputable. The very fact that the branch model has to be constructed in isolation, based on existing economic indices, and on available or accessible information in the absence of optimal decisions in other links, means that the calculations take on an approximate, conditional nature and that for the time being many problems must be solved intuitively, on the basis of expertise; this creates divergent views and a basis for debate.

Essentially there are two opposing points of view: the simplified realistic and the theoretical point of view. The first point of view (represented by I. Ia. Birman and others) consists in the realization of the above simplified model on the basis of information constructed according to the existing indices: actual or projected enterprise cost of production (or existing prices and rates), capital investments per unit of output, and planned demand coupled with the principle of minimal outlays as an optimality criterion. This naturally evokes objections since the enterprise cost of production does not correspond to the prices in the optimal plan, demand provided for in the plan is not optimal, etc. While recognizing the validity of these objections, one should not exaggerate them. First of all, certain shortcomings are partially eliminated in the model itself: to a certain extent, the consideration of the limited nature of natural reserves is tantamount to including rent in the price while the consideration of calculated outlays brings in the fund component (in particular, if both are stipulated partially for contiguous branches as well). The failure to consider commodity shortages in long-range calculations is not so essential, and the shortage of manpower can be compensated by making the proper selection of an effectiveness norm. Subsequently, the situation will improve in connection with the fact that the newly introduced prices are closer to the prices of the optimal plan than the previous ones. In part these shortcomings can be compensated by refining the model and by making adjustments in prices and norms in those cases in which such corrections are particularly significant and indisputable. In particular, it is important to understand the approximate nature of the model and not to dogmatize either its structure or the initial data or the conclusions arrived at; the latter must be subjected to additional analysis (this analysis is greatly facilitated by shadow prices obtained with the plan). Under these conditions calculations on the basis of the models yield sufficiently satisfactory results. As a rule, they are much better than those derived from the traditional consolidation of the results of the economic analysis of individual objects.

The "theoretical" approach (7) consists in the maximal approximation of the branch model to that structure of it which it receives under such a development of the optimal planning system when outlays are determined on the basis of shadow prices and production volume is determined on the basis of maximum national economic effect. The unquestionable theoretical merits of such an approach notwithstanding, it is extremely difficult to obtain satisfactory values for the necessary indices on the basis of data which are at the disposal of the branch (e.g., the calculation of the economically justified volume of use of its commodities in other branches). Moreover, the low degree of the stability of the solution with respect to changes in these indices can lead to major errors if they are not satisfactorily determined. Therefore, we must be very skeptical as to the possibility of making actual immediate use of such a model.

In addition to this general problem, there are also a number of individual debatable questions: how to treat losses incurred as the result of closing down an existing enterprise, whether to consider capital investments already made, how to treat transport, depreciation, etc. A general theoretical approach in the solution of such questions is provided by an analysis of the national economic effect under optimization conditions.

Another debatable question is the necessity of immediately establishing uniform methods for the mathematical-economic calculation of the distribution or development of a branch. In our opinion such a solution is premature. At the present time, owing to the aforementioned factors, it would be better if such models were used creatively, with a critical attitude toward the structure of the model itself and the input information, taking into account the specifics of the branch and other concrete conditions, and with an analysis being made of the conclusions arrived at and with the necessary corrections being made in these conclusions. In the face of such a situation, formal methods might only hamstring the scientist and the practical worker. Nonetheless, it would be useful to create a number of guidebooks on methods, which would explain the theoretical principles of this problem and generalize accumulated experience.

Models of Day-to-Day Planning and the Possibilities of Using Them

Even though mathematical models of day-to-day planning, as a rule, are simpler than long-range models and give promise of a more rapid practical effect, there are far fewer examples of calculations by mathematical methods of current branch production plans and their use than in the field of long-range planning. In part this is explained by the fact that there are many variants in development plans; hence greater effect is obtained from their optimization. Calculations made for certain branches (rolled metal goods, pulp-paper industry, and others) give evidence of considerable reserves for increasing productive capacity and lowering outlays on processing (on the order of 5-8 percent), which are revealed by optimal calculations. In addition to curtailing production and transport outlays, there is greater substantiation and a higher degree of reality in production plans calculated on the basis of actual productivity rather than according to consolidated indices. Moreover, there is a consolidation of orders and a corresponding additional increase in productivity, as well as better satisfaction of the consumers' orders. Moreover, in branches that have reserves of production capacities and the possibility of maneuvering them, one can expect an even greater effect when the composition of the production program is varied.

To a considerable extent, the aforementioned difference is also determined by organizational causes. While the preparation of branch development plans is concentrated in central planning institutes having the necessary time for such preparation, to a considerable extent industry was split up among the economic councils. As a result, even now there is no sufficient preparatory period in which to draft day-to-day plans. (8)

In connection with the foregoing it is necessary to pose the problem of making more widespread use of optimal models in day-to-day planning problems for a number of production branches and the problem of devising methods for this goal and the incorporation of these methods in practice. This is important for the following purposes: the obtaining of a considerable immediate effect, verification in short periods of the principles and possibilities of optimal planning, which will promote the proper assessment of these methods and their dissemination. Calculations for the determination of the structure and distribution of day-to-day production programs and the introduction of these calculations will further the accumulation of experience, of methodological approaches, and of the factual data needed to improve branch development plans. It should be borne in mind that in compiling the day-to-day plan, many problems which arise in long-range branch planning take a more concrete and discernible form: the aggregating of commodities, the evaluation of productive capacities, possible variants in the technology and organization of production, effectiveness of commodities for consumers, the evaluation of requirements and demand, consideration of transport outlays, etc. Moreover, the very posing of the problem of new capital investments in the branch becomes completely justified only after the problem of the optimal utilization of existing productive capacities has been solved. Finally, optimal calculations connected with the utilization of existing productive funds and with the evaluation of their effectiveness in the overall plan provide objective data for the construction of a number of indices needed in carrying out the economic reform, for establishing a substantiated sales plan and its distribution among enterprises, for differentiation of payments for funds and profitability norms, and for establishing the relationships of prices for various types of branch output.

It is essential that optimal day-to-day work plans for the branches and for the rational distribution of the production program be speedily worked out and put into practice. This can have a large economic effect in the near future, both in industry and agriculture. As shown by foreign experience and work carried out in the USSR at the Institute for the Economics of Agriculture (VNIIESKh) and the Siberian Department of the Academy of Sciences of the USSR, the calculation of an optimal plan for the distribution of agricultural production is completely realizable and leads to a marked improvement in economic indices. First of all, this takes the form of a scientifically substantiated definition of the optimal structures of areas under crops and of the specialization of individual agricultural enterprises as well as entire regions of the country. The effectiveness of these methods of economic analysis — particularly as applied to problems in irrigated agriculture — was demonstrated at a conference devoted to these questions, held in Novosibirsk in March 1967.

We must discuss one other problem that relates to optimal branch planning. At the present time work has begun on automated management systems for a number of branches. Unquestionably, the use of effective means of automation and mechanization of branch management and documentation flow will facilitate the solution of many problems in branch planning and economics, including the introduction of optimal programming methods. However, it would be incorrect to believe that automation of management is a preliminary and necessary prerequisite to the realization of optimal principles in the day-to-day and long-range management of a branch. As experience gained from the introduction of a number of elaborations shows, optimal calculations can be made sufficiently well and incorporated with success even before the management is automated, by using the existing means of mechanization (e.g., mechanical calculation stations) and the computer facilities of other institutions. The cause would be considerably furthered by a system of computer centers operating on a cost-accounting basis with good mathematical service. On the other hand, the development and even partial realization of optimization principles make it possible to concretize the requirements for automated management systems, to better understand their structure and the necessary technical means and flows of information. In any case, the demand that both problems necessarily be solved simultaneously and in relation to one another may do more to hinder than to promote the solution of these problems.

Models of Branch Technical and Economic Policy

In addition to the overall program for the development of the branch, a number of problems arise pertaining to the most effective system of technical solutions which on the whole determine the branch's technical policy. The selection of these solutions is inseparably bound up with their economic effectiveness and must be resolved jointly, taking into account various factors. A number of problems relating to economic policy (policies relating to prices, incentives and material self-interest, rent payments, lease payments, utilization of assets, establishment of norms for working capital and reserves, distribution of capital investments) also arise in the branch. The possibility of re-

solving these questions on a new basis — on the basis of
the methods of mathematical optimal planning — will further the intensive penetration of technology and production
by economics. Here we shall confine ourselves only to
certain questions in this area (inasmuch as questions relating to technical and economic policy are closely inter-
twined, we shall not separate them).

The structural composition of equipment. In replenishing the
equipment of existing enterprises and in planning new ones,
there arises the problem of providing for interchangeable types
of equipment (equipment of various capacities, general-purpose
and special-purpose equipment, equipment with varying degrees
of automation). The problem is particularly timely in the case
of a changing and uneven or indeterminate production program
(seasonal fluctuations, dependence on demand and receipt of
orders, etc.). These questions, for example, arise and are subjected to mathematical-economic analysis and calculation with
respect to the machine-tractor park in agriculture. The Institute of Mathematics of the Siberian Department of the Academy
of Sciences of the USSR, together with the Siberian Department
of the All-Union Institute for Mechanization, have worked out
methods to this end. These questions have also arisen in power
engineering, construction, machine building, and transport. The
solution of these problems is of great importance not only for
the branches using the given equipment but also for the branches
that produce it (determination of the range of models of tractors
to be produced, the proper proportions for manufacturing various types and models of machines, etc.). Finally, this is also
important for properly determining the economic effectiveness
of certain new types of equipment.

Technical policy in acquiring and writing off equipment, the
age structure of the machine park, depreciation payments.
Proper technical policy in the area of acquiring and writing off
equipment also plays an essential role in the case of nonuniform
equipment utilization which is typical for many branches. Mathematical analysis of the corresponding model establishes the pos-

sibility of achieving a considerable reduction in the volume of capital investments by observing a proper policy and a corresponding age structure of the park. This analysis makes it possible to define more precisely the conditions of effective capital repair and leads us to important conclusions concerning payments for funds and the structure of depreciation payments. It would be well to establish two types of the latter — payments for the calendar time and for operating time, with a differentiation according to the degree of wearout (working age) of machines. Such payments would promote the proper makeup and utilization of equipment.

This analysis also leads to the necessity for refining the calculation of the effectiveness of capital investments with respect to depreciation, and this must also be reflected in calculations of branch development models. Such an adjustment will enhance their structure, creating more favorable conditions for equipment as compared with buildings and structures. (9)

Introduction of new machinery. The introduction of new machinery requires the correct determination of the dynamics of the volume of its output and of the procedure by which it will replace previous types of machines. A proper optimal model can also be used for the solution of these important problems. This model is based on data concerning possible areas of application of the new machinery, the volume of demand for it, and the effectiveness of its use as compared with old equipment in each such area, and also on data pertaining to the possibility of producing this machinery. On the basis of this model, it is possible to calculate the dynamics of the volume of production of new machinery or of new commodities in general, the successive spheres of its utilization, the dynamics of prices and payments — depreciation, payments for funds for new and old machinery at various stages during which the old machinery is replaced by the new. In all cases we are thinking of a national economic optimum.

Since in certain periods the production of new machinery may prove to be unprofitable or only slightly profitable, it seems to us that for this period it is necessary to establish government subsidies for enterprises producing it or for the consumer. For this purpose it would be expedient to create a special fund of considerable size for subsidizing new equipment. This fund would be formed from deductions from profits obtained from the sale of those types of output that have already been mastered and that are in widespread use. The lack of such a fund places in a difficult position either the enterprise that organizes the production of new machinery or the consumer if the difference in cost is placed on his shoulders (for example, a steamship line when it receives the first ship in a new series, costing approximately twice as much as the ships subsequently produced in the same series).

Analysis of need and demand for branch output and of influences on demand. The nomenclature and structure of commodities produced by a branch must be more dynamic. They must be evaluated not so much in terms of volume (tons, rubles in fixed prices) as in terms of their effectiveness for the purchaser. To this end it is necessary that systematic studies be made of the effectiveness of a given commodity, of the need and demand for it (potential for new types of commodities). Such studies are more simple and can be made through more objective methods with respect to the means of production than the means of consumption. In the latter case, together with statistical methods and data on demand, for the purpose of evaluating use value and the dimensions of need, use can also be made of the data of other sciences (physiology and medicine, psychology) as well as sociological research — comparison of types of passenger transport, comparison of various types of dwellings and means of construction, etc. In this process it is important to determine both the personal preference and the effectiveness of various types of consumption from the point of view of society as a whole.

Need for commodities and demand must not only be studied passively. The branch must effectively influence them, stimulating an increase in orders or in demand for those types of commodities for which consumption and production are particularly effective. The following can be cited as measures which generate the necessary demand and realization of these very types of commodities: economic (prices, depreciation payments, benefits, taxes, credits), technical (servicing), propaganda and advertising, and in certain cases, administrative and planned measures. In particular, mathematical models can be used to calculate rates and payments for services that entail a more uniform demand for services in time (electric power, repair, transport) and which correspondingly result in a reduction in production outlays and capital investments.

All this indicates the necessity for making bolder use of the conclusions and proposals of economic science in practice. We must be as exacting concerning the utilization of economic progress as concerning technical progress. The administrator must bear the same responsibility for unutilized potential and lost income as he does for losses. In less indisputable problems we must resort to economic experimentation without fearing a certain element of risk.

Prices and economic indices. In addition to the overall level of prices throughout a branch — determined on the basis of national economic calculations and analysis — it is very important that there be a proper structure of prices for individual types of commodities produced by the branch. The problem of calculating such prices is connected with the simultaneous calculation of differentiated payments for funds and of norms of profitability, rents, and mining rent, taking into account interchangeability and the problem of ensuring the rational utilization of the commodities by the consumer. In approaching the solution of this problem, use may be made of models, calculating methods, and concepts of optimal planning. Here even wholesale prices and cost-accounting payments must be reduced to shadow prices in the first approximation only. In fact, however, considering the nonlinearity, the integer-valued nature, the dynamics, and the incomplete information, as well as cost-accounting independence and relations, such prices may deviate discernibly from the shadow prices. These shifts in prices must be realized in the branch by a price policy which, by means of prices and other economic levers, would make possible the optimal functioning of the economy, taking into account the conditions indicated. Moreover, in the branch price policy the total national economic effect rather than the special interests of the branch must be the decisive factor. Similar problems arise in the formation of retail prices even though the latter in principle must differ both in terms of level and of relationships from wholesale prices and must be subordinate to their own requirements and accordingly be described by other models.

In addition to this there must be much greater differentiation in prices and rates than at the present time — by season, by conditions of order and delivery, by size of orders, and in some cases, by consumer and designation. In many cases sales can be replaced by lease, rent, or temporary servicing. Such a differentiation of prices on the basis of uniform common principles must be carried out in the branch, taking into account the special features of the branch. The very methods for calculating price lists in accordance with the new structure of prices must be worked out with reference to the branch. Internal accounting prices and payments must also be formed within the branch.

The branch must develop and communicate to the consumers a scientific forecast of its development, its technology and economic indices, and in particular, the levels of prices and outlays. All these problems must also be scientifically resolved on the basis of the proper mathematical models in conjunction with statistical and normative data.

Mathematical models and general principles of optimal planning must also be applied to a considerable extent in rationally structuring branch management and its functioning. These models and principles must become a stable part of the arsenal of branch plan and economic management.

Notes

1) We consider the term "development" [razvitie] preferable to "distribution" [razmeshchenie] of the branch since the models indicated not only deal with problems relating to the distribution of new enterprises and the determination of the volumes of production but also with such problems as the size of enterprises, specialization, selection of technology and raw materials, utilization of branch output, etc.

2) See, for example, M. A. Iakovleva, "Programma dlia resheniia transportnoi zadachi," in the collection Optimal'noe planirovanie, issue 6, Novosibirsk, 1966.

3) See L. V. Kantorovich and V. L. Makarov, "Optimal'nye modeli perspektivnogo planirovaniia," in the collection Primenenie matematiki v ekonomicheskikh issledovaniiakh, vol. 3, "Mysl'" Publishers, 1965.

4) See the report by Iu. I. Volkov, "O nekotorykh uproshchennykh priemakh analiza i resheniia tselochislennykh zadach lineinogo programmirovaniia," at the All-Union Conference on the Use of Economico-Mathematical Methods and Computers in Branch Planning and Management (Novosibirsk, 1966), as well as articles by D. M. Kazakevich and others in the collection Modeli i metody optimal'nogo razvitiia i razmeshcheniia proizvodstva, Novosibirsk, 1965.

5) See L. V. Kantorovich, Ekonomicheski raschet nailuchshego ispol'zovaniia resursov (chap. 2, para. 3), USSR Academy of Sciences Press, 1959; B. Rakitskii, "Ekonomicheskie funktsii platy za resursy," Voprosy ekonomiki, 1966, no. 12.

6) See V. O. Cherniavskii, Effektivnost' ekonomicheskikh reshenii (Ocherki po voprosam sovershenstvovaniia i optimizatsii planirovaniia), "Ekonomika" Publishers, 1965.

7) See, for example, V. A. Mash, "O zadache optimal'nogo razvitiia narodnogo khoziaistva na perspektivu v otraslevom i territorial'nom razreze," Ekonomika i matematicheskie metody, 1965, no. 6.

8) It should be pointed out that Gossnab is now devoting great attention to these questions. It is essential that the production ministries also devote such attention to them, the more so in view of the fact that of late the structure and methods of calculating such models on the basis of different variants of the machine operating [stanochnaia] or distributive problem have been improved. For example, the Institute of Mathematics of the Siberian Department of the Academy of Sciences of the USSR has solved problems of this type involving 8,000 limitations and 300,000 variables (the problem of the optimal distribution of orders to rolling mills). Certain methodological problems relating to initial information have also been worked out. A number of models of day-to-day planning, connected with the distribution and specialization of production programs, have been resolved by LEMI, TsEMI [Central Economico-Mathematical Institute], and other organizations. Also of interest is the development of incompletely determined models for the distribution of the production program, which reserve part of the productive capacities, considering additional orders and changes in the structure of demand. These changes presuppose the close to optimal utilization of capacities.

9) See L. V. Kantorovich, "Amortizatsionnie otchisleniia i otsenka effekti-

166

vnosti novoi tekhniki v sisteme optimal'nogo planirovaniia," Matematiko-
ekonomicheskie problemy, Leningrad University Press, 1966; L. V. Kantorovich and I. V. Romanovskii, "Amortizatsionnye platezhi pri optimal'nom
ispol'zovanii oborudovaniia," in the collection Doklady AN SSSR, 1965, vol. 162,
pp. 1015-19.

13

Depreciation Charges and Estimation of the Effectiveness of New Technology in a System of Optimal Planning*

L. V. KANTOROVICH

The study of models of optimal planning is extremely important, not only for its own sake but also because it enables us to
determine an efficient approach to quantitative and, to a certain
extent, to qualitative study of both general and specialized problems of the socialist economy, the very nature of which corresponds to optimal planning. In our theoretical and experimental
research on planning optimization in Novosibirsk, we have also
paid a great deal of attention to such central problems as price
formation, to which the extremely interesting papers of V. S.
Nemchinov and V. V. Novozhilov have been devoted at this meeting. We also adhere to rational principles in price formation
and hold a similar but more monistic position, relating prices
to the optimal plan. We believe their direct, rather than com-
ponent-wise, calculation to be more feasible and rational. Our
viewpoint on these matters is given, for example, in [1-4]. Obviously, however, all these proposals do not yet bear a definitive
character, which they will get through further research, discussion, and experience in practical use.

Along with general issues such as price formation and the efficiency of capital investment, analysis of the optimal plan

*Matematiko-ekonomicheskie problemy, Leningrad, 1966.

provides an approach to a more efficient solution (and its objective basis) for a number of more specific economic problems. Incidentally, study of these issues also is important for the more specific and detailed study of general problems of planning and the system of economic indicators.

In this paper it is from this aspect that we consider questions of more rational size and forms of depreciation charges, evaluation of the efficiency of new technology and rational sequences of equipment replacement (which we touch quite cursorily), and the determination of prices for new equipment. (1) We have found it necessary to make our concepts about depreciation more precise in some of our specific research, such as, in calculating the optimal structure of the stock of tractors and agricultural equipment, the use of truck transport, and the calculation of rational taxi fares.

Correct calculation of depreciation costs and determination of depreciation payments is important for planning-economic analysis and cost accounting and, especially, for solving problems related to the following:

a) the evaluation of national economic costs entailed in the use of equipment;

b) the selection of the most rational use of equipment and the correct incentive structure for this selection;

c) the determination of cost of products produced with the use of this equipment; and

d) planning of the payback of costs to create and use equipment (it should be noted that under the conditions of a socialist economy it is often the case that repayment may be effected not directly but through the effects achieved in other industries).

A wide range of questions which are susceptible to analysis have still been far from completely elaborated, so that for some of them we will confine ourselves merely to the statement of the problem or an indication of the general principles of the solution. Unfortunately, we could not also include in the work concrete illustrative material for various kinds of equipment.

As we know, a number of incorrect ideas have held sway in issues concerned with depreciation (for example, concerning the role of obsolescence in a socialist economy). Many disputable and unresolved issues exist even now. It seems to us that the use of optimality principles and indicators of the optimal plan permits us to clarify many of these issues. This analysis leads to a significant reconsideration of a number of ideas and permits us to reach interesting practical conclusions.

Calculation of the costs of replacement, which is what we have in mind for the most part, is very closely connected with the calculation of the immobilization of resources in capital investments undertaken when some equipment is produced, or in other words, with the time factor. Therefore, it is very important in their calculation to adopt a methodology for estimating the effectiveness of capital investment.

As we know, the position that it is necessary when choosing capital investments to consider the time factor in the form of the norm of effectiveness has entered the methodology for calculating effectiveness adopted by the USSR State Planning Commission and the Academy of Sciences. (2) This methodology has to a certain degree an eclectic, inconsistent character. Thus, while recommending a formula for discounted value mathematically similar to the price of production when comparing alternatives, it nevertheless repudiates this formula when calculating current operating costs and relative capital investment. This inconsistency also has an effect on the calculation of depreciation charges for renovation, which usually are given in current form without discounting differential cost time streams.

In our analysis of depreciation costs in connection with the principles of optimal planning, we should use a dynamic system of shadow valuations conditioned by the optimal long-range plan. However, in certain situations these valuations for industrial products, when they are not produced through the expenditure of scarce raw materials, are close in structure to the formula for prices of production (with a single objectively conditioned norm of effectiveness)

$$c = S + aK, \tag{1}$$

where S is current operating costs; K is capital investment; and α is the norm of effectiveness. We will proceed from a structural formula such as that for national economic costs in [4]. For simplicity we will assume α to be constant and use the appropriate rule for commensurating differential cost time streams (see [2], chapter III, and, especially, pp. 184-86). As we have already noted, here we will consistently consider the depreciation costs and costs which arise as a result of the immobilization of capital investment (the time factor).

The Case of Constant Equipment Loads

We will begin with the simplest case in which obsolescence (technological progress) is not considered and it is also assumed that the operating quality of the equipment, the productivity and costs connected with its use, are unchanged over its service life. It is assumed that current and capital repair are included in operating costs. The conditions (and intensity) of utilization of the equipment are assumed to be known and constant; the service life of the equipment is n years, its acquisition cost is K rubles, and its resale value is zero.

In the comparison we will consider the commensurated cost to provide the given equipment for an unlimited (theoretically infinite) period. In this case future replacement costs, if we assume a machine which has worked for i years, are equal to

$$K_i = \frac{K}{(1+\alpha)^{n-i}} + \frac{K}{(1+\alpha)^{n-i+n}} + \frac{K}{(1+\alpha)^{n-i+2n}} + \ldots = K\frac{(1+\alpha)^i}{(1+\alpha)^n - 1}$$

(summing discounted costs on the acquisition of machines after $n-i$ years, after n more years, and so on). In particular,

$$K_0 = K\frac{1}{(1+\alpha)^n - 1} \text{ and } K_n = K + K_0.$$

It is useful to introduce the coefficient r_i for the reduction in value of a machine which has served for i years. This coefficient represents the relative reduction in the cost of replacement due to the present existence of that machine:

$$r_i = \frac{K_n - K_i}{K_n - K_0} = \frac{K_n - K_i}{K} = \frac{(1+\alpha)^n - (1+\alpha)^i}{(1+\alpha)^n - 1}.$$

We note that for small αn the expression for r_i is close to the generally assumed

$$r_i \approx \frac{n-i}{n}.$$

The depreciation charge a is based on the assumption that the depreciation charges, which under the assumption of constancy of effect are naturally assumed to be constant over the entire period, would compensate for the cost of acquiring and maintaining throughout the entire period:

$$a \cdot \sum_{j=1}^{\infty} \frac{1}{(1+\alpha)^j} = \frac{a}{\alpha} = K_n = \frac{K(1+\alpha)^n}{(1+\alpha)^n - 1},$$

where

$$a = \alpha K \frac{(1+\alpha)^n}{(1+\alpha)^n - 1} \qquad (2)$$

(if αn is small, $a \approx \frac{K}{n}$).

It is interesting to compare the size of the depreciation charges (in terms of K) for different time periods, determined by this formula, with the usual formula $a = \frac{K}{n}$ and with the value of expenditures on equipment used in calculating the commensurated cost of the product by the "Methodology for Calculating Effectiveness" according to the formula

$$a = \frac{K}{n} + \alpha K.$$

Depreciation Charges and Effectiveness

In the following table we show data for the case of $\alpha = 0.2$ (a recoupment period of 5 years). It is obvious from a comparison that the usual calculation sharply understates the replacement cost, while calculation of effectiveness according to the accepted methodology gives an exaggerated estimate of replacement costs and the cost of immobilizing capital in relation to the correct treatment of the time factor (column 3 of the table).

Comparison of Depreciation Charges by Different Methods

Service life in years	Formula for constant relative depreciation	Calculation of commensurated costs	Methodology for calculating effectiveness
1	1.0000	1.2000	1.2000
2	0.5000	0.6545	0.7000
3	0.3333	0.4748	0.5333
5	0.2000	0.3345	0.4000
7	0.1429	0.2772	0.3429
10	0.1000	0.2389	0.3000
15	0.0667	0.2139	0.2667

Here we consider the use of the equipment as a whole. If it consists of individually interchangeable components, the depreciation charges are calculated by data on the wear of individual components.

We will distinguish two factors which determine wear: time (corrosion, etc.) and the actual work time for the equipment. In other words, it is recognized that the wear of the equipment depends on the intensity of its use. It seems to us that the expended equipment may quite satisfactorily be expressed by the semiempirical formula

$$J = \gamma n + \beta t, \tag{3}$$

where n is the service life, and t is the discounted number of working hours (or other measure, such as the number of ton-kilometers or output of product, which describes the volume of operations of the equipment over the period). The total resource (through the appropriate choice of the multipliers γ and β) may be taken as equal to unity, since the equipment is useful as long as $J \leqslant 1$.

The usually applied norms which describe the physical wear either in terms of calendar time or in terms of work time are special cases of formula (3), calculated when $\beta = 0$ or $\gamma = 0$. For a given operating procedure (τ working hours a year) the service life n_0 can be defined as:

$$\gamma n_0 + \beta \tau n_0 = 1; \quad n_0 = \frac{1}{\gamma + \beta \tau}.$$

However, formula (3) permits us to consider wear and tear also for the case of using a machine following different procedures.

On the basis of an average rate of utilization of a machine and an average service life, depreciation costs (charges) may be calculated by formula (2):

$$a = \alpha K \frac{(1+\alpha)^{n_0}}{(1+\alpha)^{n_0} - 1} = \alpha K \frac{(1+\alpha)^{\frac{1}{\gamma + \beta \tau}}}{(1+\alpha)^{\frac{1}{\gamma + \beta \tau}} - 1}.$$

When the number of hours of machine use in a given year departs by $\Delta \tau$ from the average amount τ, depreciation charges must be adjusted by adding the term

$$\frac{da}{d\tau} \Delta \tau = \frac{da}{dn_0} \cdot \frac{dn_0}{d\tau} \Delta \tau = \frac{\beta n_0^2 \ln (1+\alpha)}{(1+\alpha)^{n_0} - 1} a \Delta \tau,$$

which takes account of the additional wear of the equipment. In other words, when the number of hours of use differs from τ by one hour, the depreciation charge must be raised or lowered by the amount

$$\frac{\Delta a}{\Delta \tau} = \frac{\beta n_0^2 \ln (1+\alpha)}{(1+\alpha)^{n_0} - 1} a.$$

174

If we limit ourselves to the case of $\gamma = 0$, in which wear depends only on utilization, then $\beta n_0 \tau = 1$ and the expression may be put into the form

$$\Delta a = \frac{n_0 \ln(1 + \alpha)}{(1 + \alpha)^{n_0} - 1} \cdot \frac{\Delta \tau}{\tau} = ka \frac{\Delta \tau}{\tau},$$

where the coefficient $k = \dfrac{n_0 \ln(1 + \alpha)}{(1 + \alpha)^{n_0} - 1}$ is less than unity; for small αn_0 it is close to unity. From this it follows that the incremental differential hourly depreciation charge is less than the average. It would be desirable in such circumstances to divide the depreciation charge into two parts: one proportional to time of utilization, and the other proportional to calendar time. (3) In using equipment for τ' hours, the depreciation charge will be

$$a' = a_1 + a_2 = (1 - k)a + ka \frac{\tau'}{\tau}.$$

When the number of hours of use τ' coincides with the average τ, this charge turns out to be equal to a.

Example. Suppose that $\gamma = 0$; $\beta = 0.00005$; $\tau = 4000$; $n_0 = \dfrac{1}{\beta \tau} = 5$ years; $\alpha = 0.2$. Then $a = 0.33 K$. We get

$$k = \frac{5 \ln 1.2}{(1 + 0.2)^5 - 1} \approx 0.61.$$

so that, in this case

$$a' = (1 - k)a + k \frac{\tau}{\tau} a = 0.39a + 0.61 \frac{a}{\tau} \tau',$$

i.e., under a normal load 61 percent of the depreciation charge depends on the load and 39 percent does not. This differentiation of depreciation costs and charges by use intensity of equipment derived by dividing the charge into two parts stimulates a more intensive utilization of equipment when it is desirable, and in this regard it discourages unnecessary work stoppages and inefficient utilization.

We should note that when it is difficult to calculate actual time of use of individual types of equipment, this time should be calculated by means of data on the output of products, by dividing the depreciation charge into one part that does not depend on the volume of output and one part that is proportional to the volume of output calculated on the basis of normative machine time requirements per unit of each type of product. However, in this form these incentive measures may be used only during operating regimes that do not differ significantly from the average. When there is a major difference between operating regimes, the question of depreciation charges cannot be resolved in isolation from the question of the most rational use and distribution of equipment.

The Case of Unequal Machine Load

Actually, a given piece of equipment may be applied in different conditions and with a different intensity. In this case the question of the optimal sequence for acquiring it and for using equipment which has already been acquired must be resolved by methods of optimal programming. In order to give an idea of how this will affect depreciation costs, we will consider the question in real concrete conditions where such an optimal distribution is determined in a natural way, since the analysis of even this relatively simple case presents undoubted interest.

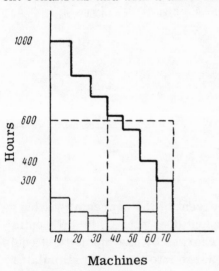

Distribution of Load

176

Depreciation Charges and Effectiveness

The situation in which the quantity of necessary equipment is determined by peak loads — the necessity for providing for the periods of greatest demand — is quite normal. At the same time, in other periods only a part of the existing equipment may be used (tractors, taxis, machines, energy equipment). In other words, the load curve looks like that shown in the diagram. For example, the greatest load for the 70 machines lasts for only 300 hours; only 60 machines are necessary for a period of 100 hours, etc. The fine line shows the distribution of loads, and the heavy line the total number of hours during which the load will be at least equal to that indicated on the horizontal axis. We will denote the average number of hours per machine by τ (the broken line in the diagram). Then if for simplicity we assume that $\gamma = 0$, the service life of one machine is equal to

$$n_0 = \frac{1}{\beta\tau} \; .$$

We will assume that the load levels $a_1 \leqslant a_2 \leqslant \ldots \leqslant a_r$ are multiples of the same number, or more precisely, $a_i = iM$. Suppose that the time during which the load is not less than a_i is equal to τ_i. Obviously,

$$\tau_1 \geqslant \tau_2 \geqslant \ldots \geqslant \tau_r > 0.$$

The optimal schedule for acquiring and operating machines is as follows. The total number of machines rM (which provides for the peak loads) is divided into r equal parts, and the i-th part operates during a period of $\frac{n_0}{r}$ years with a load in $\tau_i \frac{n_0}{r}$ hours. (We assume that time is continuous and does not require that the identical regime be maintained throughout the the entire year.) For each $\frac{n_0}{r}$ years we acquire M new machines, M worn machines are written off, and the remaining machines are transferred to the group with lower intensity of operation (from the i-th to the $i + 1$-th).

If at the starting point we have M machines that have operated for $\dfrac{n_0}{r} \displaystyle\sum_{j=1}^{i-1} \tau_j$ working hours $(i = 0, \ldots, r-1)$, then the result-ing schedule is stationary and the given state of the process repeats with period $\dfrac{n_0}{r}$ years. If it is assumed that in cases in which the initial state of the process does not satisfy the proportionality conditions, the process nevertheless (if M is sufficiently large) leads to a stable proportional state.

We will now consider the example of an even simpler case in which $M = 1$, $r = 2$. In this case we assume that

$$\frac{1}{(1+\alpha)^{\frac{n}{2}}} > \frac{\tau_2}{\tau_1 + \tau_2}$$

Depreciation charges for a machine which has operated for r hours amount to $(\underline{4})$

With	For Δt years of service	For $\Delta\tau$ hours of work time
$r \leqslant \dfrac{n\tau_1}{2}$	$\left\{ \tau_1 \left[(1+\alpha)^{\frac{n}{2}} - (1+\alpha)^{\frac{r}{\tau_1}} \right] + \tau_2 \left[(1+\alpha)^n - (1+\alpha)^{\frac{n}{2}} \right] \right\} S\Delta t$	$(1+\alpha)^{\frac{r}{\tau_1}} S\Delta\tau$
$r > \dfrac{n\tau_1}{2}$	$\left[(1+\alpha)^n - (1+\alpha)^{\frac{r}{\tau_2} + \frac{n}{2} - \frac{n}{2} \cdot \frac{\tau_2}{\tau_1}} \right] \times \tau_2 S\Delta t$	$(1+\alpha)^{\frac{r}{\tau_2} + \frac{n}{2} - \frac{n}{2} \cdot \frac{\tau_1}{\tau_2}} S\Delta\tau$

where
$$S = \frac{K \ln(1+\alpha)}{\left[(1+\alpha)^{\frac{n}{2}} - 1 \right] \left[(1+\alpha)^{\frac{n}{2}} \tau_2 + \tau_1 \right]}.$$

Analysis of differential costs that arise in performing operations for these or other periods with the required optimal stock leads to the following, at first glance unexpected, results:

1. Depreciation costs and charges are uniquely divided into charges for calendar time of utilization and charges for work time.

2. The size of these and other charges are differentiated according to machine category and according to degree of wear (period of service). For depreciated machines the calendar time charge must be much lower, and the work hour charge higher, than for new machines. The ratio of the charges is such that they optimize the use of machines (under a rational stock). In particular, they justify the using of worn machines only during the peak periods. At the same time, because of the low calendar time charges for these machines, the enterprises will not be interested in an accelerated write-off.

Today single depreciation norms and single planning indicators, in agriculture for example, make the retention of old machines for peak periods undesirable and sometimes provoke premature conversion to equipment that exceeds demand during peak periods.

3. The depreciation charge for work time must be differentiated by periods depending on the load. It must be reduced for the off-peak loads (today the reverse is frequently done).

4. In determining the time cost we should develop and apply more differentiated and detailed calculations of depreciation charges. Mechanization of these calculations makes differentiated calculations quite realistic.

5. Calculation of depreciation costs as a function of the aforementioned principles leads to substantial changes in the effectiveness of capital investment.

We will briefly consider the possibilities for taking into account a number of other influences on the calculation of depreciation costs.

For a number of machines there is a reduction in productivity, an increase in operating costs (materials and fuel), an increase in repair costs, and so on. If we have the necessary data, these can all be taken into account. This shows the need for differentiation of depreciation and redistribution of depreciation cost and charges, i.e., for some increase in them during early periods and a reduction later on.

Consideration of obsolescence of the first kind (reduction of the cost of production of the particular equipment) already leads to a redistribution of depreciation costs and an increase in their share during the early years of the service life of the machine. The same may be said concerning obsolescence of the second kind (i.e., consideration of technological progress, the appearance of new versions of this type of equipment with better features) to the extent that it can be anticipated (specifically or generally).

We also note that when there is a difference between the volume and composition of the actual equipment stock and the optimal stock, we can introduce adjustments into the charges connected, in particular, to the scarcity or excess supply of one or another type of equipment. Generally, the prior determination of the size of depreciation charges must be adjusted for technological progress and the state and development of the economy.

During the course of expanded reproduction, improvements in technology, and technological progress there is a continual replacement of equipment. In this connection it becomes necessary to resolve a number of economic problems concerning the desirability of the sequence and stages for replacing old equipment by new equipment, the level of production of one or another kind of equipment in different periods, the correct applications and utilization of new equipment, and the level of prices and depreciation charges for earlier equipment and new equipment in different periods.

It seems to us that analysis of the optimal plan for replacing earlier equipment by new equipment enables us to find a satisfactory and consistent solution for the total range of problems that relate to the determination of the time for cutting back the production of old equipment, putting together a plan for the development of production of new equipment, determining the time for converting to the new equipment in different areas, and determining the time for complete cessation of the use of earlier equipment and for setting the price levels and depreciation charges for old and new equipment for all periods.

Such an analysis, which relies on a dynamic model of optimal planning, must correctly consider the possibility of producing one or another kind of equipment and its use at different enterprises, as well as the economic effectiveness of the use of some kinds of equipment under different conditions. This permits us to consider the price and depreciation charge levels appropriate to stimulate the optimal utilization and replacement sequence and to make profitable the production of new equipment and its application in those areas where it is rational.

It seems to us that the problems which have been raised here may become an interesting and necessary object of research at the general theoretical level as well as within the framework of individual industries.

Notes

1) E. A. Bulavskii and a number of other co-workers of the Mathematical-Economics Department of the Institute of Mathematics of the USSR Academy of Sciences, Siberian Section, participated in this research. Also see [5-7].

2) In this connection we would like to acknowledge the major services of V. V. Novozhilov for his development of and insistence on correct concepts for calculating the effectiveness of capital investment in the socialist economy and, especially, the role of the optimality principle in the methodology for estimating effectiveness. Unfortunately, in its theoretical aspect this methodology still far from fully utilizes and realizes even the recommendations in Novozhilov's prewar work on this score.

3) It should be noted that this differentiation of charges by calendar time and work time is also desirable when establishing norms for different types of operating costs, repairs, and replacement of individual components and machine parts with short service life. We also note that this differentiation is also necessary when $\gamma = 0$, since in this case it reflects the immobilization of capital.

4) This model is considered in more detail in [6] and [7].

References

1. L. V. Kantorovich, "Ob ischislenii proizvodstvennykh zatrat," Voprosy ekonomiki, 1960, no. 1.

2. L. V. Kantorovich, Ekonomicheskii raschet nailuchshego ispol'zovaniia resursov, Moscow, USSR Academy of Sciences Press, 1959.

3. L. V. Kantorovich, "Dinamicheskaia model' optimal'nogo planirovaniia," Planirovanie i ekonomiko-matematicheskie metody, "Nauka" Publishers, 1964.

4. L. V. Kantorovich and V. L. Makarov, "Matematicheskaia model' perspektivnogo planirovaniia," Primenenie matematiki v ekonomicheskikh issledovaniiakh, vol. 1, "Mysl'" Publishers, 1965.

5. L. V. Kantorovich and V. A. Bulavskii, "Vozmozhnosti ispol'zovaniia mate-

maticheskogo optimal'nogo planirovaniia v analize effektivnosti novoi tekhniki," Trudy soveshchaniia VISKhOM po effektivnosti novoi sel'skokhoziaistvennoi tekhniki, Moscow, 1965.

6. L. V. Kantorovich and I. V. Romanovskii, "Amortizatsionnye platezhi pri optimal'nom ispol'zovanii oborudovaniia," DAN SSSR, 1965.

7. L. V. Kantorovich and I. V. Romanovskii, "Struktura amortizatsionnykh platezhei pri ravnomernoi zagruzke mashinnogo parka," DAN SSSR, 1966.

14

Economic Problems of
Scientific-Technical Progress*

L. V. KANTOROVICH

Scientific and technical progress is the basic means for further increasing the effectiveness of social production in the age of developed socialism. The level of science, the significance of new scientific and technical developments and inventions, and the rate of their adoption in production and their diffusion determine a country's economic potential. Science becomes a direct productive force and the decisive factor in the development of a socialist economy.

The espenditures of the national economy on science and technology also grow immeasurably and become comparable with basic items in the national economic expenditures. For this reason, the development of the national economy is in large measure determined by the economics and management of scientific and technical progress. How can the most intensive development of science and technology be secured? What resources should be allocated for this purpose? In which directions can they be used most effectively? What attainments of science and technology should be used in what volume, and in what sequence in the national economy, in order to achieve the highest results in assuring the country's industrial and defensive

*Ekonomika i matematicheskie metody, 1974, No. 3.

potential and in raising the population's standard of living? To a considerable degree the future successes of our economy depend on the correct solution of these problems.

Even though the socialist system creates exceptionally favorable conditions for the development of technical progress and for its utilization in the national economy, the realization of these advantages is influenced by the quality of planning and of technical-economic calculations and the system of economic indicators. In order to achieve "the organic combination of the attainments of the scientific and technological revolution with the advantages of the socialist economic system" [1], we must draw upon the attainments not only of the natural and technical sciences but equally of the social sciences as well, and especially of economic science.

In this regard we cannot fail to note the significant increase in the degree of activity in economic research that has taken place in the last decade, the closer proximity of economic theory to the vital demands of economic practice, and its enrichment with new methods associated with the application of mathematics and cybernetics. New scientific cadres have grown up with a modern level of training. There have also been great changes in economic practice associated with the economic reform, and in particular, the inclusion of capital intensiveness in cost accounting and in prices has been introduced. Measures have been taken to improve management, planning, supply, and price formation, which are also based to some degree on the new economic research.

However, many economists and managers have still hardly mastered and systematically applied the new methods. At the same time, it is especially important to apply these methods to problems pertaining to scientific and technical progress. A significant and not infrequently dominant place in economic practice belongs to traditional methods in planning and in the construction of economic indicators. But while purely balance-sheet methods, planning on the basis of the attained level, and the application of averaged indicators produce results that are acceptable to some degree in questions pertaining to established types of products and gradual inertial economic development, they prove to be insufficient when we turn to economic problems of scientific-technical progress, to the economic appraisal of the production and diffusion of new products, basically new production methods, and new sources of raw materials, and to questions characterized by multiple variants and dynamism.

The system of indicators and the economic measures should create a favorable climate for scientific-technical progress and should assure interest in promoting such progress at all levels in a short period of time and in forms that are most effective at a given time and in a given place. Of great importance for the correct solution of problems of technical progress is the introduction in planning and economic practice of the attainments and conclusions of the Soviet school of optimal planning and functioning of the economy, which we shall discuss below.

However, the economic problems that arise in connection with the scientific and technological revolution are so specific that in their analysis we cannot confine ourselves to the application of already elaborated methodology. This methodology must be substantially developed, and a number of serious special studies must be made.

Scientific-Technical Progress and Problems of Economic Management

High rates of scientific and technical progress significantly complicate the economic management problem and require the enrichment of management methods so that they will take these new conditions into account. If we are discussing prospective planning — medium-term and especially long-term planning — then, taking the optimal dynamic model of such planning as the base [2], we must have the following initial data: the initial state of the economy, labor and natural resources for the period under review, the normative technology matrix, and the structure of final consumption. Under the conditions of intensive scientific and technical progress, however, it is extremely difficult to construct all these data for the future (with the exception of comparatively stable demographic data), especially since it is necessary to consider in some measure the same data for the postplan period in order to appraise the final state of the economy, which is also included in the optimality criterion. The foregoing applies in particular to the matrix of technical coefficients in which totally new, unforeseen methods may appear. Technical progress also complicates the forecast of natural resources and leads to marked changes in the structure of final consumption. In this connection the role of forecasting in long-term planning [3] mounts, especially for more extended periods. The indeterminacy and instability of the initial data require the development of methods of long-term planning calculations, e.g., the use of various levels of aggregation for different periods and the application of the methods of stochastic programming. In accordance with the principles of continuous planning, the solution calculated by means of these methods is of a final nature only for the first part of the planned period (essentially the first five-year period) and during the remaining period must be made more precisely in accordance with the new data, which have changed primarily as a result of scientific-technical progress.

The intensification of scientific-technical progress imposes growing demands not only on the selection of economic planning decisions but also on a certain flexibility of the economic system and its capacity to alter its structure effectively in accordance with the new conditions. From this point of view one must, for example, give preference (with an equal or even somewhat lesser degree of economic effectiveness) to general-purpose equipment instead of narrowly specialized equipment, since the former will find application under altered conditions as well.

The simplest one-product models can be used for the global analysis of the development of an economic system and for taking into account the influence of technical progress on the dynamics of an economic system. The influence of technical progress on the most important economic characteristics can be studied on the basis of these models.

Problems of Scientific-Technical Progress

Let us examine an economic system that produces one product, one part of which goes for consumption and the other for increasing fixed capital and working capital, which, moreover, are not differentiated in the given model.

On the assumption of the instantaneous transformability of capital and the accounting of technical progress, such a model is described by the following equation

$$\frac{dK(t)}{dt} = P(t) - V(t) = e^{\delta t}U[K(t), T(t)] - V[t, T(t), K(t), P(t)], \quad (1)$$

where $P(t)$ is the quantity of net output or national income; $V(t)$ is total consumption per unit of time. Production function $U[K(t), T(t)]$ characterizes the quantity of net output that can be produced in a unit of time assuming capital $K(t)$ and labor resources $T(t)$, and δ characterizes the rate of neutral technical progress.

An important parameter of the economic system — the norm of effectiveness of capital investments $\eta_{\mathfrak{s}}$ — can be determined within the framework of this model. For $\eta_{\mathfrak{s}}$ we obtain the expression

$$\eta_{\mathfrak{s}} = \frac{\dfrac{1}{P}\dfrac{dP}{dt} - \dfrac{T'}{T} - \delta}{1 - \dfrac{V}{P} - \dfrac{T'}{T}\dfrac{K}{P}}. \quad (2)$$

All values in this formula have a clear economic content. In particular, the model can be used to study the influence of technical progress on $\eta_{\mathfrak{s}}$ and other characteristics of the system.

More realistic is an analogous model without the assumption of the instantaneous transformability of capital [4]. It is assumed that there is a spectrum of capital of different structure. In the process of economic development there occur the release of labor resources employed in production with low capital per worker, or in production using previously created capital, and the transfer of these labor resources to newly created capital.

The maximization of the increase in the net product at any moment in time (differential optimization) is taken as the criterion. The model is described by the following system of equations

$$\varphi(t) = T'(t) + \varphi[m(t)] m'(t), \tag{3}$$

$$\varphi(t)\lambda(t) = \varkappa(t), \tag{4}$$

$$\varphi(t) U[\lambda(t), 1] - \varkappa(t) U_\lambda'[\lambda(t), 1] - e^{\delta[m(t)-t]}\varphi(t) U[\lambda[m(t)], 1] = 0, \tag{5}$$

where $\lambda(t)$ is capital per worker of new capacities put into production; $\varphi(t)$ is the intensiveness of activation of new capital; $m(t)$ characterizes the policy of withdrawing capital from production. These are the functions desired. Parameter δ characterizes technical progress. The following expression was derived for the norm of effectiveness of capital investments in this model:

$$\eta_\vartheta = \frac{1}{\varkappa(t)} \left\{ \frac{dP(t)}{dt} - e^{\delta m(t)} U[\lambda[m(t), 1]] T'(t) \right\}. \tag{6}$$

The solution of the system (3)–(5) for the concrete type of functions of $U[x, y]$, $T(t)$, and $\varkappa(t)$ makes it possible to study the dependence of η_ϑ and some other characteristics on the parameters of the system.

As a result of the large volume of spending on science and technology and their introduction into production, it is no longer possible to view technical progress as an exogenous factor in long-term planning. It becomes natural to consider the relationship between changes in the technology matrix the result of the development of science and technology and the volume of expenditures for these purposes. There must be corresponding structural complications of models of the long-term plan all the way up to the formulation of the problem of finding the optimal share of resources to be allocated for science. Even though the existing information hardly makes this feasible at the present time, the very formulation of such questions is significant [5].

Significance and Prospects of the Optimal Approach
to the Economic Analysis of Scientific-Technical Progress

Economic measures must assure maximally favorable conditions for scientific and technical progress and interest in its development and must promote the optimal management of such progress.

Therefore, at the present time the economy must be based on more precise and differentiated economic indicators and more flexible planning decisions. Valuable conclusions concerning the construction of such a system of indicators can be obtained by using mathematical models of optimal planning.

Let us enumerate those problems in which in-depth economic analysis holds particular importance for scientific-technical progress.

The introduction of payments for capital and the inclusion of capital intensiveness in price in the process of the economic reform promote the optimal utilization of new capital and the correct distribution of the product. This is not enough, however. The application of a more differentiated system of payment for capital, with due regard not only to the balance-sheet value of capital but also to the economic (rental) appraisal of capital [6], and the calculation of depreciation norms that take into account the age, degree of utilization, and obsolescence of equipment are scientifically justified. This is important for the elaboration of a progressive policy on the replacement of equipment. The economically substantiated higher level of depreciation allowances and of payments for capital during the first years of operation of equipment would promote the more intensive utilization of new types of equipment. The existing practice of distributing depreciation for renovation equally over the years without taking into account intertemporal differences places equipment (the active part) in a less favorable position compared with plant when the effectiveness of capital investments is appraised [7].

The use of depreciation allowances can become a very important means in the regulation of technical progress. Unfortunately, the results of the research in this area are as yet virtually un-realized. At the same time, quite evident is the expediency of making a radical change in the very structure of payments, which must consist of the part that is porportional to the life-time and the part that is proportional to the use-time of equip-ment [8]–[9]. Such a structure of the depreciation norm and of payments for capital can play a large part in the moderniza-tion of technical equipment.

Such differentiation is also important in price formation. The inclusion of capital intensiveness in price was a significant as-pect of the price revision of 1967. This measure, the necessity of which was substantiated by the theory of optimal planning, substantially promoted the establishment of a more correct ratio of sectoral price levels, eliminated a large number of enterprises operating at a planned loss, and improved the qual-ity of economic valuations.

Owing to the short time limit, however, the accounting of capital intensiveness has not been sufficiently consistent. As a rule, price reflects the capital intensiveness not of a specific product but of the branch, and it is precisely on the basis of this capital intensiveness that the normative profitability (vis-à-vis the enterprise cost of production) is determined on a differenti-ated basis from branch to branch.

As a result, material-intensive production is more profitable and labor-intensive production is less profitable. As a rule, in machine building the material intensiveness of old, familiar products is higher than the branch average, while the nominal calculated capital intensiveness far exceeds actual capital in-tensiveness. For this reason, old machine-building products are highly profitable. On the other hand, under the existing methodology of calculations, progressive products that are in the stage of development are relatively unprofitable. The pro-duction of these products is economically disadvantageous.

The introduction of the results of scientific-technical prog-ress is an extended process. For this reason the accounting of the time factor and the correct comparison of expenditures in-curred at different times with the results are of decisive importance here. For example, in order to determine whether the expenditures required for the acceleration of the realization of a new technological process or the production of a new product are justified, it is essential to compare expenditures and effects for a number of years.

At the same time, accounting of the time factor in practice is not of a systematic nature and frequently is entirely absent. Consistent accounting of the time factor in construction logically demands that the sum of expenditures incurred at different times be discounted [6]. Under this procedure construction organizations would have an interest (other things being equal) in the maximum reduction of construction delays and in the concentration of effort, since otherwise they would not be able to stay within their estimated costs. Therefore, if the time factor were correctly and completely considered, there would be a drastic reduction in construction delays and a smaller volume of resources frozen in incomplete construction. The resources released in the process would make it possible to accelerate the equipment renovation rate.

Consistent accounting of the time factor in the extractive industry would result in the more intensive working of deposits.

The correct comparison of expenditures incurred at different times is especially essential in appraising the economic effectiveness of scientific research. For a number of questions associated with technical progress, simplified accounting of the time factor through a single discounting coefficient based on averaged data proves to be insufficient. It is essential to consider the change in the relationship of optimal valuations of various ingredients (raw materials, machines, labor) in time brought about by differences in the rates of their decline over time. For this reason it would be more accurate to discount expenditures and effects to a single point in time by using differentiated coefficients for every group of expenditures [2], [10]. For example, if we take into account the growth of the relative valuation of labor, this will increase the valuation of the effectiveness of mechanization and automation.

Finally, accounting of the time factor is essential for the

correct economic appraisal of the quality of products designated for use in production, since such appraisal must include the results of the use of the products during their entire service life.

The intensification of production is a key direction for technical progress in agriculture. Of great importance for attaining intensification of production is the reflection of land rent in prices on agricultural products and in financial relations with the government. At the present time, even though the existence of rent under socialism is theoretically recognized, it is not properly considered in economic calculations. At the same time, systematic utilization of this category in prices, cost accounting, and relations between agricultural enterprises and the government would promote a further rise in the level of their specialization and the development of economic relations. Rent relations would particularly promote the intensification of agricultural production. Indeed, the additional output obtained from the same area of land would not be subject to rent payments, as a result of which it would not only cover production expenditures but would also be highly profitable to the economy. The decision of the April (1965) Plenum of the Central Committee of the Communist Party of the Soviet Union on the establishment of additional payment for production in excess of the plan was already directed toward the intensification of agricultural production, so that the aforementioned measures should be viewed as a natural further development of these decisions.

The establishment of more correct economic relations and indicators in agriculture will permit more precise appraisal of the effectiveness of progressive measures and will create favorable conditions for their realization [11]. It would be expedient to conduct such an experiment in one or two republics or regions of the RFSFR.

The systematic and correct accounting of mine rent in economic analysis is also of great importance. It is important for price formation in the extractive industry, for the rational working of deposits, and for the effectiveness of geological prospecting work. The introduction of the concept of valuation of mineral reserves into economic practice would eliminate the unjustified losses that result from the incomplete extraction of ore.

The rational distribution of labor resources plays an important part where technical progress and the development of new branches and enterprises are concerned. It is essential to make a correct determination — together with wages — of the social valuation of the effectiveness of various types of labor. These two indicators make it possible to introduce differentiated payments by enterprises for the use of certain categories of labor resources. This point is repeatedly discussed in the press. The use of these indicators would assure the supply of the best labor resources to new production facilities, which is a condition of no little importance to technical progress. (1) In our view there is also justification for raising the level of wages of workers and management at enterprises with relatively high capital per worker, with high-quality natural resources, or that produce new types of products, since the optimal selection and retention of cadres and high intensiveness of labor increase the output-capital ratio and also promote the formation of new production facilities.

Also important for the realization of technical progress is further work on improving the system of economic indicators, particularly the introduction of the volume of net output as a volumetric production indicator together with the sales indicator.

For economic levers — prices, payments, etc. — to be effective, they must not only be correctly established. It is also important that the system of cost accounting and the appraisal of the activity of economic objects be real and that the ruble be meaningful to enterprises. But it is not infrequently the case that not only our suppliers but, paradoxical as it may be, our customers as well are interested in high prices, and there are unjustifiably profitable and unprofitable types of products.

Only the insufficient effectiveness of the ruble can explain the fact that over a number of years hundreds of millions of tons of coal have been extracted at a cost of 12-15 rubles a ton, and only 20-30 million tons have been extracted at a cost of 1-1.5 rubles a ton.

Retail prices are of great importance. Now they are not based on a unified, scientifically substantiated system. Retail prices have formed historically as a result of a number of measures that were put into effect in different years and that were not always coordinated. Compared with wholesale prices, the levels of retail prices on different commodities of the same social significance differ several times. Some prices are unjustifiably high, others are outdated, and there are other disparities. For example, it is entirely justified that prices on fruit are several times lower in southern regions than in northern regions owing to differences in natural conditions. It would also be correct to establish relatively low prices on domestic electric power in Siberia, since electric power there is produced at a low cost. It should be said that the substantiated lowering of prices leads not only to a higher living standard but frequently may also increase government revenues at the same time. An example of this is the new taxi fares that were established at one point at the suggestion of mathematicians.

Thus the regularization of the system of retail prices on consumer goods and services can produce a significant rise in the standard of living, progressive changes in consumer demand, and an increase in trade turnover, thereby creating favorable conditions for the development of new industries.

Unlike capitalism, in our country such indicators as profit and income are not an end in themselves but are merely a means of increasing the effectiveness of the economy. The national economic effect and the interests of all society are the final goal and the criterion. In some cases this effect cannot be fully taken into account in cost accounting and in profits, and this is in the nature of things (profit does not originate, for example, as a result of the construction of highways, even though it is very important and justified). The latter circumstance must be taken into account in appraising the effectiveness of technical innovations and scientific elaborations. The essential points here are the effect realized by the customer, the influence on the future development of other branches, etc. For this reason the methods of taking full account of the national economic effect must be elaborated and put into practice. Enterprises and agencies do not always have sufficient economic incentive for introducing new equipment. This incentive must be supplemented by effective control methods. For this reason it is expedient to direct efforts toward the creation of a methodology for appraising the technical level and rates of technical progress and to elaborate a system of measures for monitoring technical development in individual branches.

From the standpoint of the national economic effect, incremental or differential expenditures on increasing output [6], [7], [12], determined by the increment in current and capital outlays, have much greater significance than average expenditures, on which prices and rates are for the most part now based. The use of differential expenditures, e.g., for the creation of a new system of rates in rail transport, would, in our opinion, alter the approach to the solution of the problem of siting production and in particular would long ago have destroyed the myth that it is uneconomical to ship Kansk-Achinsk coal, etc. Such an approach would permit the more correct determination of the rational volume of shipping and the substantiation of the effectiveness of additional investments in the development of the railroad network. Such investments would be more than compensated by the reduction in expenditures in industry as a result of the greater degree of specialization and concentration and the use of better, even if more distant, deposits.

Based on differences between differential and average expenditures in the manufacturing industry, and especially in the machine-building industry, it is possible to substantiate the reduction in prices on many types of products and the expediency of increasing the size of series of machine tools and instruments. This creates prerequisites for accelerating technical progress, for increasing the active share of expenditures (on equipment) in capital investments, and for reducing the volume of repair work.

A characteristic feature and great advantage of the socialist economy is that its main goal is the final national economic effect and not the results of the activities of individual agencies. Therefore, even if the recommended lowering of railroad rates and prices on machine-building products reduces the income level of these agencies somewhat but provides a considerably larger saving in capital construction and in other branches, it will prove to be expedient.

Cost accounting, which has undergone further development in the course of the economic reform, should be coordinated with the plan and should stimulate its fulfillment. A rationally constructed system of economic indicators and levers also has another meaning: it provides the means for making operational corrections in the plan in accordance with newly revealed needs, resources, etc., in the most effective direction.

In our opinion, under these conditions it is expedient to use less rigid forms of plan targets. In planning the production program, especially for new or recently adopted types of products, it is difficult to indicate with sufficient accuracy the possible volume of these products. In order to guarantee 100 percent fulfillment of the plan, the enterprise strives to understate the plan and to conceal its reserves and potential. Both the supplying and the organization of production are planned on the basis of this understated target. This in turn hinders the use of reserves arising in the process of implementation of the plans, and especially the increase in the production of the necessary products. Thus work performed by USSR Gossnab on the optimal utilization of pipe-rolling mills revealed the possibility for the additional production of approximately one hundred thousand tons of pipe. However, it proved to be very difficult to realize this possibility owing to the insufficient flexibility of planning.

If in accordance with the indeterminate (stochastic) character of production data the plan were organized not on a strictly deterministic basis but with due regard to expected (probable) results, perhaps, within certain limits, we could achieve a considerably larger output. Of course, such a planning procedure would also require a more flexible system of supply and availability of reserves, making it possible to correct deliveries in the course of the fulfillment of the plan. This would encourage initiative in the organization of new production. It seems expedient to verify this system of planning in an economic experiment.

The establishment of greater flexibility in plans and the expansion of the possibility for their justified correction would create conditions that are particularly favorable to technical progress, to new production, and to technical elaborations and scientific experimentation, which would make it possible to reduce the time of their realization and to accelerate the process of introducing new technology. In particular, instead of the prolonged and primitive process of "transfer of know-how" to other enterprises, we would sooner achieve large-scale standardized production of the equipment necessary for the diffusion of work methods that have proven themselves to all interested agencies.

Lead time can be reduced, for example, through the creation of firms that serve as a kind of incubator for inventions. After receiving an invention from the author, these firms would bring it to the stage at which its usefulness, competitiveness, and realizability would be evident and at which it would be possible to make an approximate appraisal of its economic effect. The introduction of the invention would subsequently be assigned to an appropriate branch. This kind of firm can operate on a cost-accounting basis, and even if only one out of every ten inventions is used, it would justify itself. This circumstance, as well as the fact that all stages would be concentrated in one place, would greatly reduce the time involved in the development and introduction of new technology.

Of course, the application of modern methods of economic analysis is also important in appraising the effectiveness of inventions.

In order to secure the most rapid organization of series production of new items created both as a result of inventions and of planned R & D, it is expedient to increase the number of enterprises that are specialized according to the technological rather than the object principle. Such enterprises may be extradepartmental or may be subordinate to a special general-purpose machine-building agency. By accepting orders from different agencies they could assure higher productivity and higher work quality, could have better specialized equipment, and could use it more intensively.

Thus a high level of economic analysis and further improvement in the organization and management of production, based on the modern means and potential of economic science, are an essential condition for the attainment of the necessary rate of technical progress.

Difficulties and Particular Features of Economic Appraisal of Technical Progress

Of the greatest importance for the solution of economic problems of scientific and technical progress is the creation of such conditions that managers, in L. I. Brezhnev's colorful phrase, would not shun new technology as the devil shuns incense but rather could expect that justified expenditures on the measures necessary for the creation of new technology would be compensated and that initiative and effort would be properly appraised.

In order to gain a better understanding of the difficulties associated with the economic characterization of new technology and ways of overcoming them, it is useful above all to note certain features of new technology. First, the process entailed in the full realization of new research, development, and inventions is an extended process, consisting of a number of stages and lasting 5-10 and sometimes even 25 or more years. For example, in the USA the time needed to create important and generally acknowledged innovations has been as follows: the production of aluminum — 31 + 6 = 37 years; radio broadcasting — 17 + 9 = 26 years; synthetic fibers — 6 + 3 = 9 years; computers — 15 + 6 = 21 years; atomic power stations — 11 + 3 = 14 years; integrated circuits — 2 + 3 = 5 years. (2)

There is also a great lapse of time between the beginning of industrial production of a new item and its diffusion in considerable volume. Thus the volume of the production of aluminum made up 0.02 percent of the national income only after a lapse of 14 years and 0.2 percent after a lapse of 67 years. For air transport these periods were 8 and 16 years, respectively; for synthetic fibers — 2 and 22 years; television — 2 and 4 years [13, pp. 227-31]. In determining the fate of new technology (a new product, new production process) and in creating economic conditions for its development, it is essential to consider in some measure this entire single, prolonged process.

The second characteristic feature of new equipment lies in the extremely dynamic character of its production and, in particular, in the sharp change in economic indicators, especially during the first years of adoption. In order to illustrate this point let us cite general data obtained at the All-Union Institute of Agricultural Machine Building by A. P. Kolotushkina on the production of a number of new agricultural machines [14] (see Tables 1 and 2).

Table 1

Coefficient of Change in Own Expenditures on the
Production of Agricultural Machinery Depending
on Year of Adoption Assuming Stable Production

Year of adoption	1	2	3	4	5	6	7
Base coefficient	1	0.80	0.72	0.67	0.65	0.63	0.62

Table 2

Scale of Change of Own Expenditures on the Production
of Agricultural Machinery Depending on Production Volume

Annual scale of production (thousands of units)	1.0	2.0	5.0	10.0	25.0	50.0
Coefficient of change	1.394	1.348	1.225	1.0	0.661	0.608

Thus, for example, if in the second year of adoption own expenditures per machine, assuming the production of 5,000 machines, amount to 3,000 rubles while the enterprise cost of production, together with purchased materials and items, is 3,800 rubles, in the fifth year — assuming the production of 25,000 machines — own expenditures will amount to

$$3,000 \cdot \frac{0.65}{0.80} \cdot \frac{0.661}{1.225} = 1,300 \text{ rubles,}$$

while the enterprise cost of production will be 2,100 rubles instead of 3,800 rubles. In the case of capital expenditures the difference is usually even greater.

We have examined changes in one short stage in the development of production. In the extended process of diffusion of new equipment, expenditures may decrease tens of times, while operational and use qualities may sharply improve.

The third characteristic feature of the production of new equipment lies in the indeterminacy of the data and in the existence of risk. Thus less than 10 percent of all inventions are applied in industry.

All these features, which play the greatest role when we discuss basically new and promising discoveries or inventions, must be taken into account in the process of resolving planning and economic questions relating to the introduction of new equipment. This refers to the decision-making process for the elaboration and industrial adoption of new equipment, the rate and volume of development of its production, the appraisal of economic effectiveness in different stages, the establishment of prices on new equipment, sources for the financing of its production, and the provision of cost-accounting conditions and material incentives.

It would seem that new technology can be appraised in economic terms by using the dynamic model of optimal planning, which in principle makes it possible to appraise the effectiveness of the use of each new method during an extended time interval: it is sufficient to calculate total expenditures and effects for the entire period according to the dynamic system of valuations, and the positive character of the result will demonstrate its effectiveness. What is more, by comparing the optimal plan compiled without the use of this method with the plan using this method (on an optimal scale), on the basis of the final results it would be possible to obtain an economic appraisal of a given discovery or invention for the economic system as a whole. New technology, however, is characterized by discreteness and nonlinearity. It can be realized in a certain minimum volume, and moreover, the expenditures and effects depend on this volume. Of course, it is possible to adopt a certain program for the development of new technology and to reformulate the optimal plan, incorporating this program in the plan. Then it would be possible to obtain an appraisal of the effect of new technology (albeit incomplete) under the selected program governing its introduction. It is difficult, however, to consider even this plan realistic, since owing to the aforementioned features, it is practically impossible to gain reliable data on economic indicators characterizing the production of new products, even more so for quite an extended period.

For the same reasons, more particular dynamic models of the diffusion of new equipment and of the calculation of expenditures and prices on such equipment are not sufficiently realistic, even though they are theoretically substantiated. An example is the model proposed by the author (together with V. A. Bulavskii) at a session of the All-Union Institute of Agricultural Machine Building in 1964 or the model developed in the work by N. Ia. Petrakov and K. G. Gofman [15].

Thus the latter models examine the question of establishing the level of prices on new equipment between the upper limit, which is characterized by the effect to the customer or the price of the equivalent base product, and the lower limit, which is determined by expenditures on new equipment. At the same time, it may develop that the lower limit during a certain period of time will be higher than the upper limit; nonetheless, the production of new equipment will be expedient, and the losses will subsequently be recovered with interest when its production process is improved and it comes into common use. However, the introduction of the requirement of an immediate and certain economic gain from new equipment would make it necessary to abandon this equipment in some cases, while in other cases it would impede the diffusion of this equipment for a long time. Moreover, this could more frequently be found specifically with respect to basically new equipment, especially that which is effective in the future.

Since new equipment and the conditions governing its production radically differ from the production of old items, economic analysis of questions pertaining to new equipment must differ from the general methods used in appraising the effectiveness of economic measures. It has already been stated that it is also impossible to directly apply the system of valuations of the optimal plan to this analysis.

In view of this, taking the particular features of new equipment into account, it seems more expedient to us to use another forecasting approach to its appraisal.

The impossibility of obtaining any reliable data concerning future indicators of the production of new products, especially for an extended period, or of precisely calculating economic results on their basis, compels us to base our analysis on approximate, conditional data found by means of forecasting methods on the basis of the statistical generalization of existing economic experience from new products and of various expert appraisals.

It is primarily essential to characterize the expenditures and results that arise in new production. In addition to the conventional type of expenditures that are also found in the case of old products (fixed and working productive capital, labor expenditures, expenditures on technological design, etc.), new production entails a whole series of other types of expenditures. It requires:

    a) extensive preliminary research and development work;

    b) fabrication and testing of prototypes;

    c) the elaboration of a production method and, in certain cases, the design and fabrication of new equipment as well;

d) losses associated with the interruption of already mastered production or with the lowered volume of this production during the period of conversion and mastering of new production;

e) expenditures on personnel training;

f) expenditures associated with the risk of failure (the production of a new product may prove to be economically ineffective, or the technical and economic parameters of a new product may fail to correspond to the design parameters); and

g) expenditures associated with learning how to use a new product (change in the production method, equipment and maintenance, information, training, etc.).

At the same time, in case of success the organization of new production entails a number of benefits for the national economy that are in no way confined to the direct material effect embodied in the actually produced new product. Economically speaking, this effect is only partially reflected in the income derived from its sale. For this reason it is also essential to consider a number of other elements of the national economic effect derived from new production. Let us note some of them:

a) the real effect received by the customer for a new product frequently exceeds its price (moreover, by different amounts for different categories of customers);

b) since it is more modern, the equipment that is used for new production (newly created or modernized) usually also has a higher economic valuation;

c) in the case of new products compared with old products, the process of improving designs and production methods is more rapid, the output volume increases more rapidly, and sales are guaranteed to a greater extent (on the whole this creates potential conditions for the growth of future income);

d) designs and production methods elaborated while producing the first series of products reduce expenditures and improve product quality in subsequent production and at other enterprises mastering the production of these products;

e) the mastering of new production (or of a new production method) often entails progressive changes in such related industries as the production of analogous types of products, supplies, and equipment required for a given or similar type of production. This engenders new needs and new technical requirements and enriches the directions of research and development realized in a given product by practical experience.

In short, for a certain period of time only a part of the economic effect derived from the adoption and production of a new product will be directly embodied in the product. The other, sometimes no smaller, part consists in increasing the general economic and scientific-technical potential as a result of mastering new production.

While the specific types of expenditures that originate in the production of a new product are quite well known and recognized, the great long-run benefits that the new product yields are usually not taken into account in quantitative terms, even though they are qualitatively understood. However, the ignoring of these benefits in economic calculations means an incorrect orientation in questions pertaining to the appraisal of new technology, places it in a disadvantageous position, and is a hindrance to technical progress.

Consideration of this "aftereffect" of new production indicates that even if the production of a new product proves to be unprofitable — expenditures or losses exceed gross income — in the ex post or planned interval, this does not mean that the organization of its production was unjustified. It is entirely possible that the potential benefits from new production will more than compensate for the losses that are sustained.

This seemingly paradoxical conclusion, that new products and new production methods yield an immediate economic effect only in a few instances but are unprofitable in the initial stages in a considerable number of instances (3), should not frighten us or incline us toward abandoning their production. In creating an appropriate system of economic measures and valuations it is necessary as far as possible to consider the full national economic effect derived from the use of new technology. Thus the approach to the economic substantiation of new technology must basically differ from calculations of the effectiveness of conventional economic measures. Without attempting to offer finished proposals on this question, we shall merely note some of the considerations that in our view are most important.

Planned and ex post data on new equipment can actually be constructed only for a limited period of mastering the production of this equipment and for the first years of production. It is impossible to make planning calculations for a longer period of time, and the ex post calculations are more of historical interest. However, as has been shown, the postplan period, the aftereffect, must not be ignored. The effect of this period must be considered by forecasting methods using generalized statistical data on new equipment as well as branch features and certain expert estimates, e.g., the probable realizability of an invention, the possible volume of diffusion of a given new product, or the volume of application of a new production method. A typical example of this type of estimate is the above-cited generalized coefficients of lowering of expenditures on agricultural machines as production develops and volume increases. It is possible to create and use special norms of this type for individual types of new products.

Thus in the process of making calculations pertaining to innovations, we must in one way or another take into account the postplan period, the dynamics of development of production, and the characteristic economic features of this period, as well as the indirect results of the diffusion of the innovation.

Certain Conclusions and Proposals

We shall briefly describe the way in which such an approach should affect the analysis of the effectiveness of new equipment and the creation of economic conditions for its realization. This is especially important when we are concerned with basically new equipment rather than with partial modernization or new parameters.

Making decisions on the production of new equipment. According to available data, we must appraise the necessary expenditures (capital and current) by the time of starting up of new production. On their basis we must determine calculated expenditures per unit of new product and compare them with the price of the base product or the effect of the new product.

Considerable corrections must be made in these calculations. If we are concerned not with the one-time production of a new product but rather with its systematic diffusion in the national economy, of decisive importance are long-run costs of production attained when the diffusion process is completed rather than the actual expenditures that form during its initial production or, in general, at the first enterprise. These expenditures should rather serve as merely one of the guideposts for making a more realistic appraisal of long-run costs of production.

For this reason, in the calculations there is every justification for not considering the part of capital expenditures that is associated with higher costs of producing the first models of the equipment, a large part of expenditures on research and development that can justifiably be imputed to subsequent production, as well as part of the expenditures on the reorganization of production. A correction may be made for the lower volume of production stemming from the use of nonspecialized equipment, and the reduction of expenditures in the process of the development of production can also be taken into account. As a rule, such corrections can be made not through concrete analysis but rather on the basis of certain norms for individual types of expenditures.

The long-run calculated expenditures derived in this way should be compared with the effect of the product, or with expenditures on the base product that it replaces, for the purpose of resolving the question of the effectiveness of new equipment and the justification for its production.

The economic effect resulting from the creation of new equipment. This effect must be appraised in different stages of the development and production of new equipment. Here it is possible to make a preliminary, long-run valuation based on the probable volume of its diffusion, on the expected level of expenditures per unit of the new product, and correspondingly on the effect derived from it (differentiated by sphere of application if possible and discounted to a single moment).

The calculation of the economic effect can also be based on results obtained in the process of organizing the production of new equipment in the planning stage or after adoption. In such a case the economic effect of its utilization can be calculated directly. In this kind of calculation, however, only part of the actual expenditures should be included. The other part should be imputed to future outputs and to the productive contribution to technical progress (norms can be established for the admissible magnitude of this share for capital expenditures and expenditures on fabrication). It is expedient to consider not only the effect derived from products already produced but also from future production in the immediate years to come, when the production costs will be lower. If we are concerned with a basically new product, and if at the time of the evaluation the scale of diffusion of the product or its influence on related branches are clear, these effects can be added to the previously determined effect.

It should be said that even with such a just and simultaneously stimulating valuation of the effect, by no means will all measures pertaining to new technology prove to be justified and yield an effect. New ventures entail inevitable risk. The efficiency of a new product must be determined not by the rigid demand that each individual type of product or new production method pay for itself but by the efficiency of the technical policy governing the elaboration and industrial adoption of innovations as a whole.

Failures and unjustified expenditures in the production of some items are permissible if they are more than compensated by the effect derived from the rapid and successful realization of others. Of course, the allusion is to failures due to objective obstacles and difficulties and not to unconscientious work.

Financing and cost accounting. As had been stated, the production of new products, even though it is justified and ultimately yields a positive national economic effect, may prove to be unprofitable for a number of years, and the output and income may not correspond to the expenditures that have been incurred.

Clearly, under such conditions an enterprise producing a new product will find itself in a difficult situation for no good reason unless special measures are taken.

A similar situation obtains when we make capital investments. In such a case the expenditures are also recovered only after a number of years. But special resources are allocated for this purpose from the accumulation fund, and no one demands that they be returned from the output of the first two or three years.

Accordingly, the process of organizing the production of a new product (even at an existing enterprise) should also be regarded as a long-term measure requiring considerable expenditures, and its effect should be appraised with respect to the entire production-development process.

For the same reason as in the case of capital investments, this process cannot be realized on the basis of the enterprise's own current resources (development fund) and, as a rule, must be financed by allocations from centralized funds (branch or national economic). Moreover, such allocations must be made over an extended period of time, not only during the period of adoption but during the first years of development of production as well.

Therefore, along with the fund for the development of new technology, which should be considerably increased by including in the enterprise cost of production of old products, let us say, 5 percent of the payment for obsolescence, there would probably also be justification for using some portion of the resources from the government or branch capital investment fund for this purpose. This is all the more natural since the conversion of an enterprise to the production of new, more valuable products means the modernization of an enterprise, leads to an increase in the economic effectiveness of its fixed capital, and thereby compensates for capital investments.

Certain other incentive measures are also possible, e.g., exemptions from capital charges for several years, and perhaps from depreciation allowances as well for enterprises mastering the production of new products.

Under these conditions cost accounting can be organized in

such a way that enterprises adopting and producing new products will not find themselves under worse conditions than enterprises producing old products. Material incentives for new equipment, including incentives for inventors, must be provided accordingly. Since the valuation of the effectiveness of new equipment is made with successive corrections in several stages of its development and adoption in production, it is also expedient to offer the incentives in stages.

Prices on new products. If we have a new product the production of which is justified, then the conditions of its production during the adoption period, during its first years, or even at the pioneer enterprise in general are such that the price on this product, calculated in the conventional way on the basis of expenditures, will not only not correspond to long-run (socially necessary) expenditures on it but may prove to be significantly higher than its effect to the customer or the price on the product that it replaces. Such a situation hinders the correct use and diffusion of the product, is disorienting with respect to the area of its application in the planning stage, and delays its diffusion. On the other hand, the establishment of a lower price will make a new item totally or relatively unprofitable and will place its producer in a disadvantageous position. What is the solution? It appears expedient to proceed from the same thesis, that part of the expenditures incurred in the initial production should be imputed not to the specific product but to the very creation of the new type of equipment, to the raising of scientific-technical potential. Under these conditions the price to the customer should be established on the basis of long-run costs of production, including normal profitability, and consideration should also be given to the benefits to the consumer from the new product. This will promote its broadest possible diffusion and effective use. As for the enterprise, it is expedient to establish an accounting price that will cover its substantiated expenditures and assure the profitability norm adopted in the branch, or even a higher profitability (up to 50 percent). The difference between the accounting price and the transfer price must be covered by the development fund (or the technical progress fund).

It seems especially important that the problems of planning new equipment, of appraising its economic effectiveness, and of finances, cost accounting, and prices be resolved not in isolation from one another but in mutual coordination, with due regard to the specific features of new equipment.

We have presented certain principles for the approach to the solution of these problems, albeit in very sketchy form.

The objection has been raised that such a solution, which takes into account the specific features and difficulties of new equipment, creates hothouse conditions for it and impedes the correct selection of technical variants.

It seems to us that such an approach is just as necessary as it is to grow seedlings in hothouses if we wish to have an early harvest. On the other hand, the ignoring of the specific features of new equipment in economic analysis compels us either to abandon it entirely or frequently to artificially adjust the data in order to justify its effectiveness under the existing methods.

Thus the aggregate of scientifically substantiated economic incentives for new products and of constraining measures for obsolete products is an important means for the acceleration of technical progress. It is to be expected that such measures will prove to be more effective than administrative decisions.

Notes

1) It is interesting to note that such payments exist — and on a large scale — in Great Britain, for example.

2) The first figure denotes preparatory work, and the second figure indicates the organization of industrial production.

3) Of course, if an innovation is justified, then by calculating its production and diffusion as well as its aftereffect for an entire period of 20-30 years, we would find that the useful effect and the income would far outweigh the costs of production. However, we do not have data at our disposal to make an economic calculation for such a period of time. Cost accounting does not extend over such a long period of time and is even less applicable to the effects in other branches. Therefore, we have to reckon with the fact that a positive balance is not obtained during the realistic period of time.

References

1. L. I. Brezhnev, Otchetnyi doklad Tsentral'nogo Komiteta KPSS XXIV s"ezdu Kommunisticheskoi partii Sovetskogo Soiuza. Moscow, Politizdat, 1971.

2. L. V. Kantorovich and V. L. Makarov, "Optimal'nye modeli perspektivnogo planirovaniia," in Primenenie matematiki v ekonomicheskikh issledovaniiakh, vol. 3. Moscow, "Mysl'" Publishers, 1965.

3. N. P. Fedorenko, O razrabotke sistemy optimal'nogo funktsionirovaniia ekonomiki. Moscow, "Nauka" Publishers, 1968.

4. L. V. Kantorovich and V. I. Zhiianov, "Odnoproduktovaia dinamicheskaia model' ekonomiki, uchityvaiushchaia izmenenie struktury fondov pri nalichii tekhnicheskogo progressa," Doklady AN SSSR, 1973, vol. 211, no. 6.

5. B. S. Mitiagin, "Zametki po matematicheskoi ekonomike," Uspekhi matematicheskikh nauk, 1962, vol. 27, no. 3 (165).

6. L. V. Kantorovich, Ekonomicheskii raschet nailuchshego ispol'zovaniia resursov. Moscow, USSR Academy of Sciences Press, 1959.

7. A. L. Lur'e, Ekonomicheskii analiz modelei planirovaniia sotsialisticheskogo khoziaistva. Moscow, "Nauka" Publishers, 1972.

8. C. Abraham and A. Thomas, Microeconomic Optimal Decision. Paris, Dunod, 1966.

9. L. V. Kantorovich, "Struktura amortizatsionnykh platezhei v nekotorykh modeliakh ispol'zovaniia mashinnogo parka," in Optimizatsiia, issue 4(21). Novosibirsk, 1971 (Institute of Mathematics, Siberian Branch of the USSR Academy of Sciences).

10. L. V. Kantorovich, V. N. Bogachev, and V. L. Makarov, "Ob otsenke effektivnosti kapital'nykh zatrat," Ekonomika i matematicheskie metody, 1970, vol. VI, no. 6.

11. L. V. Kantorovich and M. I. Virchenko, "Matematiko-ekonomicheskii analiz planovykh reshenii i ekonomicheskie usloviia ikh realizatsii," Voprosy analiza planovykh reshenii v sel'skom khoziaistve, part 1. Novosibirsk, 1971 (Siberian Branch of the USSR Academy of Sciences).

12. V. V. Novozhilov, Problemy izmereniia zatrat i rezul'taty pri optimal'nom planirovanii. Moscow, "Nauka" Publishers, 1972.

13. E. Mensfild, Ekonomika nauchno-tekhnicheskogo progressa. Moscow, "Progress" Publishers, 1970.

14. Metodika opredeleniia optovykh tsen na novye sel'skokhoziaistvennye mashiny. Moscow, Preiskurantgiz, 1969.

15. N. Ia. Petrakov and K. G. Gofman, Rekomendatsii po ustanovleniiu tsen na novye vidy produktsii s pomoshch'iu elementarnykh optimizatsionnykh raschetov. Moscow, 1968 (Central Economic-Mathematical Institute of the USSR Academy of Sciences).

Concerning Prices, Rates, and
Economic Effectiveness*

L. V. KANTOROVICH

Mathematics is a powerful instrument that is used in the search for optimal solutions to economic problems. The importance of elaborating and introducing economic-mathematical methods has been emphasized in the decree of the Central Committee of the Communist Party of the Soviet Union on the development of the social sciences and in recommendations of the all-union conference on improving planning and economic work in our country. At the present time the results of economic-mathematical research are being widely used by USSR Gosplan. It is not without reason that a significant place is devoted to these results in the recently published methodological guidelines on the compilation of the national economic plan. (1) Mathematics and computers are used in the elaboration of plans for the development of many branches and in resolving all manner of economic and technical problems.

As a rule the introduction of mathematical methods assures a great economic effect to the national economy. Thus, for example, as a result of the new taxi fares recommended by mathematicians, the population benefited by more than half a billion

*Ekonomika i organizatsiia promyshlennogo proizvodstva, 1971, no. 1.

rubles in five years time, while the government realized almost the same amount due to the curtailment of deadheading and idle time. I can also refer to such works performed by the Institute of Mathematics of the Siberian Branch of the USSR Academy of Sciences in conjunction with other institutes as the determination of the optimal utilization of the country's rolling and pipe mills, the calculation of the optimal location of agricultural production in various natural-climatic zones of the Omsk Region, etc.

Under the new conditions for the management of the national economy, at a time when the role of economic methods of management is emphasized and cost accounting is being introduced and improved in every possible way, the significance of optimal planning increases even more.

The ideas of optimal planning have already found some embodiment in the process of the economic reform. In particular, in the 1967 price review the capital and, in part, the rent component are considered in the structure of prices. Further measures for improving price formation are being elaborated by the State Committee on Prices of the USSR Council of Ministers, with the participation of specialists on optimal planning. Mathematical methods permit a more substantiated approach to the ascertainment of economic indicators, to the system of material incentives, and to the calculation of rent payments and other indicators that are necessary for improving planning and management.

Optimal planning has its own methodology for constructing economic indicators, and hence one must approach critically many economic and statistical data and the conclusions based on these data.

Shortcomings of Price Formation

At a conference on the development and location of Siberia's productive forces, held in May 1969, it was reported that, according to statistical data, labor productivity in Siberia was

approximately 15 percent higher than in the European part of
the country. However, the labor productivity indicator, as it is
presently calculated, is very arbitrary, since it takes into account not only value added but also transferred value, turnover
tax, etc. If we are guided by this indicator, we can conclude,
for example, that the population should for the most part be concentrated in regions with a high concentration of the confec-
tionery, sewing, and spirit-vodka industry, since labor productivity in these branches is much higher than in the mining and
coal industry.... But if we consider the particular features of
price formation that determine the unduly low valuation of the
products of the extractive and certain other branches that make
up the basic share in the structure of Siberia's industry, labor
productivity in Siberia will prove to be not 15 percent higher
but much more so, and arguments in favor of the preferential
development of Siberia will become even more convincing.

The incorrect relationship of prices on the products of various branches of industry stems from the fact that price formation — especially in the mining and timber branches — does not
take sufficient account of differential rent, and as a result the
effect and advantages of using rich natural resources are not
fully disclosed. While Kuznetsk coal is on a par with Donbas
coal, the price of the former is almost one-half that of the
latter....

The correct accounting of rent holds especially great importance for agriculture. Such accounting would lead to significant
changes in prices on agricultural products and in financial relationships between agricultural enterprises and the government.
At the present time differences in the natural fertility of the land
are allowed for through prices: the level of purchase prices is
differentiated according to agricultural zones, and differential
rent is collected in proportion to the volume of production. If
procurement prices are made approximately uniform throughout the nation and are appreciably raised in the majority of regions, and if fixed rent payments for natural resources are established, this will promote the fuller and more effective use of
the land. The best and most fertile plots will be used more

intensively, production on them will increase, and the incomes of collective and state farms will be higher. At the same time, the government's expenditures associated with the rise in the purchase prices would be exceeded by significant revenues in the form of rent payment and income tax. When uniform prices are elaborated, they need not be immediately applied throughout the entire country. Instead, we should conduct an economic experiment in one region or republic and satisfy ourselves as to the expediency of instituting such prices. The time is evidently ripe for such an experiment. (2)

The difference in prices (for example, the prices on coal) from one region to another is sometimes attributed to high shipping costs. However, we cannot rely on the existing railroad rates and shipping costs in these instances, since the calculations employ average indicators rather than concrete indicators for a given type of shipment and a given sector in the network. This difference may be very considerable. For example, the average shipping cost (for 1,000 ton/km) over the Omsk-Novosibirsk trunk line is 0.90 rubles, compared with the average network cost of 2.30 rubles.

The transport branch (especially rail transport) is characterized by the fact that incremental expenditures on freight turnover (both current and capital) are 2 to 3 times lower than average expenditures per ton/kilometer. That is why real labor productivity in rail transport increases so rapidly and shipping costs keep falling. Thus between 1960 and 1965 rail freight turnover increased on the average by 6 percent at the same time that there was a certain reduction in the work force, while shipping costs fell by 3.5 percent a year. Of decisive importance in economic calculations are the incremental expenditures, which differ greatly from average expenditures in this area. It is also necessary to consider the fact that with a twofold increase in the distance of shipping (let us say from 700-800 to 1,500 kilometers) the shipping costs increase by only 1.5 to 1.6 times. The rates in certain countries, e.g., in the USA, are constructed accordingly. In our country the kilometer rates are practically uniform, while in the past we even had prohibitive rates that increased with distance.

The idea which we expressed at one time as an intuitive hypothesis (3), concerning the importance of taking into account the differentiation of expenditures according to types of shipment and regions, has been fully confirmed by detailed analysis using factual data. (4)

The refined analysis of real shipping costs leads to the reexamination of many questions pertaining to territorial location: rational output volume, geographic zones of attraction of enterprises, basins, deposits, etc. The doubt arises as to whether it is advisable in many instances to duplicate production in new places while production capacities are not being completely utilized at existing enterprises (for example, building materials enterprises); whether it is expedient to discontinue long-distance shipping of consumer goods, etc. Of course, the development of transport and the increase in the volume of shipping require certain expenditures of material and monetary resources, but in such cases they are significantly exceeded by the saving in current and capital expenditures in industry and construction.

From our point of view, for example, the existing opinion that it is not economical to ship Kansk-Achinsk coal is by no means indisputable. Actual events have destroyed the myth of the technical impossibility of shipping this coal over long distances, and yet another myth — the myth that the long-distance shipping of this coal is uneconomical — must also be destroyed. The actual cost of shipping this coal will probably be not 1 ruble 75 kopeks (the average network cost of coal shipments) and not 3 rubles (the rate for shipping coal) but approximately 70 kopeks per 1,000 ton/kilometers, i.e., the shipment of this coal to the Urals and even to the Center is entirely economical (national economic expenditures per ton of standard fuel will be not more than 5 to 8 rubles). And since this is the case, the zone of use of Siberian coal, and especially of Kansk-Achinsk coal, can be considerably expanded and the rate of its production can be increased. There is clearly no justification for extracting hundreds of millions of tons of coal for 14 to 16 rubles a ton while at the same time extracting negligible quantities of coal costing 1 ruble a ton.

The deviation from real expenditures is also characteristic of passenger rates, especially in air transport. There is every economic justification for introducing air transport rates that go down with distance (similar to railroad rates). The population would benefit and government revenues would also grow, since the shipping costs in air transport would decline as a result of the increase in volume. Contrary to common sense, the existing rates encourage long-distance travel by rail and short-distance travel by air.

It would be expedient to introduce reduced rates for individual types of travel (medical-connected travel, vacation travel) as well as seasonal rates. Such an approach would have great social significance. It would help to bring the living standard of the population of Siberia closer to the living standard in the European part of the USSR and would offer a psychological inducement for people to take jobs in Siberia.

Taking into account real expenditures in the establishment of prices and rates leads to the conclusion that residents of Siberia should use electric power at a reduced rate since the cost of producing it in Siberia is significantly lower, just as southerners pay less for fruit. The differentiation of rates for electric power not only by region but also by season, day of the week, etc., would be one of the factors stimulating the better use of the fixed capital in the power industry and would offer an economic incentive to enterprises to attain a more even consumption schedule. In particular, rates established according to this principle would promote the wide use of electric power in agriculture. Then it may become advantageous to make use not only of fixed irrigation facilities for vegetable and technical crops but also of mobile and contingency irrigation facilities for grain crops, which is of great importance in drought years in Siberia and Kazakhstan. Power is needed for irrigation in the summer, i.e., in the period of underutilization of electric power stations. If a reduced rate is established for agricultural enterprises during this period, the consumption of electric power will increase, and the resulting effect will exceed the losses associated with the lowering of the rate.

Rational Utilization of Labor Resources

The social assessment of labor used in economic-mathematical calculations characterizes the magnitude of the effect that can be provided by an increase under given conditions in labor resources in the economy. This assessment does not coincide with the wage indicator. For example, it is possible that an increase in the wage fund by 100,000 rubles would make it possible for an enterprise to increase its net output by 300,000 rubles by hiring additional manpower. Then in the given instance the 300,000 rubles will be the real assessment of the indicated increase in labor resources. This assessment plays an important part in the comparison of the effectiveness of utilization of labor resources in one or another region as well as of various categories of labor. Even though it is widely used in planning practice, the wage indicator does not reflect the total expenditures that are associated with the utilization and hiring of labor resources. As a result there is insufficient economic incentive for releasing manpower through the introduction of new equipment, especially in regions with a shortage of manpower, and for the proper distribution of labor resources among branches.

In the interest of making rational use of labor resources, especially in cities and regions with a manpower shortage, it is expedient to introduce a special system of payments by enterprises for the use of labor as a whole or for individual categories (for example, the labor of young males). (5) Such payments would stimulate a more even employment of the population from one region to another, more effective utilization of labor, a high level of automation and of the availability of fixed capital, and better supply of labor resources to the most important industries. The funds received could be used for measures to eliminate the shortage of labor resources, e.g., to improve living conditions in remote regions of the nation, especially in Siberia. As we know, in some countries — Great Britain, for example — such labor-use tax amounts to a considerable sum at enterprises in individual branches: the total revenues from this tax exceed all government expenditures on education and health.

The Time Factor

Frequently, in our concern for our descendants, we act to the detriment of the present generation, and it may turn out that our sacrifices are in fact unnecessary. This applies in particular to the use of mineral deposits. For example, it is still the dominant conviction that mines should necessarily be operated for a long period of time, in any event, for a period at least equal to the amortization time of the basic equipment. The period of working a coal deposit is set at least at 40 to 45 years, and the period of working a gas deposit is set at 20 years. In 20 years, and all the more so in 40, we will unquestionably have considerably cheaper means of producing energy, as is evidenced by the progress already made in atomic energy. What is more, at the present rate of technical progress equipment becomes obsolete quite rapidly.

Moreover, under the existing approach to the assessment of the effectiveness of increasing the annual extraction of fuel and reducing the time that deposits are worked, the actual effect is frequently underassessed.

Let us imagine that at the same time that we reduce the period of working of a deposit from 40 to 20 years, we can increase the output of a mine from 3 to 6 million tons a year. Many believe that the resulting lowering of the enterprise cost of production, e.g., by approximately 15 percent, is not so very great. In reality, however, the mine of doubled capacity will not only replace the projected mine but will be the equivalent of yet another mine with an output of 3 million tons, at which the enterprise cost of production and capital investments per unit of output will be 30 percent lower than at the first mine. If we were to build this second mine in another place, the expenditures on it would be not lower but would probably be even higher than those at the first mine (unless an equally favorable field were found). Such reasoning also holds true for oil and gas deposits. In the USA, for example, it is the practice to work lenticular (local) deposits of oil and gas for 5 to 6 years.

The intensive working of deposits will sharply increase the extraction of minerals in the next few years, and the losses associated with the incomplete amortization of the equipment will be offset. Thus there must be a high degree of dynamism in the working of deposits, and the period of their working must be shortened. It is no less important to cut the time entailed in the construction and development of deposits. We can even permit a significant increase in one-time capital investments if this makes it possible to cut construction time in half.

The correct assessment of expenditures on construction projects is important for speeding up construction. This assessment must be made not on the basis of the simple summation of estimated expenditures but must be formed as calculated cost (taking into account the timing of expenditures on construction and on the acquisition of equipment and their discounting to the point at which a project begins operation). It is also important to institute large penalties (up to 50-100 percent of the cost of deliveries) for the violation of deadlines for the delivery of supplies and equipment.

Finally, correct accounting of the time factor, changes in the magnitude and structure of depreciation allowances, and scientifically substantiated differentiation of payments for capital would promote more complete utilization of equipment and would demonstrate the economic advantages of investments in the active part of equipment compared with investments in buildings and structures.

The intensification of the utilization of existing structures is a problem that also merits special attention. The specialized Leningrad organization of Giprospetsgaz, for example, has analyzed variants of pipeline operation. The intensification of

the Bukhara-Ural gas pipeline through the installation of additional compressor stations has been examined. It has been calculated that a twofold increase in the number of compressor
stations (with the necessary technical elaborations) would make
it possible to increase the carrying capacity of the gas pipeline
by one-third and to obtain an additional 8 to 10 billion cubic
meters of gas a year without the expenditure of pipe. For the
intensification of the utilization of gas pipelines, it is also possible to make use of additional units at compressor stations,
including those that have been installed as backup units. In
connection with the installation of additional units, of course,
it will be necessary to increase their production; but our
machine-building industry is equal to the task, and it is even
profitable for the industry, since the unit enterprise cost of
production falls sharply as the volume of series production
increases.

* * *

The USSR national economy has now entered a stage in which
the intensification of production is increasingly becoming the
most important direction in its development. The use of modern methods of economic science — particularly the ideas of
optimal planning — will promote the attainment of the objectives that have been set.

Notes

1) Metodicheskie ukazaniia k sostavleniiu gosudarstvennogo plana razvitiia narodnogo khoziaistva SSSR, Moscow, "Ekonomika" Publishers, 1969.

2) For greater detail, see L. V. Kantorovich, "Puti primeneniia matemati-cheskikh metodov v sel'skokhoziaistvennom proizvodstve," in Optimal'nye mo-deli orosheniia, Moscow, 1969.

3) L. V. Kantorovich, Matematicheskie optimal'nye modeli v planirovanii razvitiia otrasli i v tekhnicheskoi politike, Novosibirsk, 1966. Also see Voprosy ekonomiki, 1967, no. 10, and Kommunist, 1966, no. 10.

4) A. I. Zhuravel', "Issledovanie vliianiia dal'nosti perevozok gruzov na tekhniko-ekonomicheskie pokazateli zheleznykh dorog," Trudy NIIZHTa, no. 80, Novosibirsk, 1968.

5) See L. V. Kantorovich, Ekonomicheskii raschet nailuchshego ispol'zovaniia resursov, Moscow, 1959, chap. II, para. 4.

16

Ways to Develop Computing Means for Solving Large Optimal Planning and Control Problems*

L. V. KANTOROVICH

The solution of large optimal planning and control problems requires very cumbersome computational operations that are based on a very large body of initial data and that require the large volume of data processing entailed in the application of modern mathematical methods. One encounters a large volume of computations both in precise methods (linear algebra, linear programming, etc.) and in iterative, gradient, and especially combinatorial methods, in the methods of scanning, discrete and stochastic programming, and simulation methods. The complexity of the problems and the volume of the computations increase considerably with the transition from static to dynamic and forecasting problems.

The mechanical means and programming languages used at the present time are not well adapted to the solution of problems of such a massive character. Even though the universality of machines and languages makes it possible to obtain programs and solve problems, the structure of these machines and languages, based on operating with individual numbers, does not make the application of these means sufficiently effective and hinders their use. In our opinion the basic factors that impede

---

*Optimizatsiia, 1972, no. 6.

the introduction of modern methods for solving large problems lie in the following:

1. Operating with individual units of information greatly complicates algorithmic descriptions and programs and makes them inflexible and inadaptable to the necessary variations and combinations, to the varied use of information, etc.

2. Programs prepared in this fashion, and especially programs obtained through translation, lead to the uneconomical use of technical means. Owing to the existence of a large number of auxiliary commands (preparatory and control commands), the working time of the machine is not expended economically and productivity is lowered. Nor is the machine's memory used with sufficient efficiency, and this lowers its productivity even further.

These difficulties are especially important in the case of operations that are within limits close to the resolving capacity of the machines, which leads to the excessive demands on the power of the machines.

It seems to us that these shortcomings are to a considerable degree connected with the universal character of the languages and machines, which does not take into account the structure and other features of the large problems that we have been discussing.

Among the important features of these problems, we should list:

1) The existence of information blocks organized in a certain way and, as a rule, used in a certain way.

2) The fact that the major part of mathematical algorithms in these problems essentially lies in the application of certain group operations to the indicated information blocks.

Taking these features into account, it should be possible to obtain much more effective means for the solution of large problems.

To this end we consider it important to coordinate the effort of specialists in various professions in this direction: economists, mathematicians, programmers, and engineers to assure a systems approach to the solution of the problem of providing effective software and hardware for these important and complex calculations.

The systems approach should lead to the creation of new machines and languages that will be effective for the solution of the aforementioned economic problems by their specific mathematical algorithms and that will permit a convenient description of these problems and their translation into effective computer programs.

We believe that the importance and mass character of the aforementioned problems justify the elaboration of computer complexes primarily oriented toward the solution of these problems. It seems to us, however, that such complexes will not be narrowly specialized and will probably prove to be effective for many other classes of problems as well.

In the process of preparing for the solution of the problem posed, it is expedient to perform the following preparatory work:

1. To classify and standardize current and long-term planning and control problems.

2. To standardize methods and algorithms broadly used at the present time and in the future for the solution of the given range of problems (from the area of optimal programming, control theory, theory of games, linear algebra, operations research, statistics, and information theory).

3. To analyze characteristic methods and algorithms from the standpoint of the structure of initial and intermediate information blocks, and forms of their organization and processing, and to analyze mass conveyer, cyclical, and other operations that are the most time consuming.

4. To study technical possibilities and technical means for the effective realization of operations on information blocks and for the effective organization of the transmission and processing of these information blocks.

5. To ascertain the principles involved in constructing a high-level language that permits the convenient and effective description of the indicated methods, that is oriented toward the corresponding technical means, and that takes their possibilities into account.

6. To ascertain the principles of translation and interpretation of problems described by such a language and to make effective use of consolidated information units and operations made available by technical means.

17

A Modern Mathematical System
of Economic Management*

L. V. KANTOROVICH, E. G. GOL'SHTEIN,
V. L. MAKAROV, and I. V. ROMANOVSKII

The problem of improving planning and management has acquired decisive importance for increasing the effectiveness of social production. This is evidenced, in particular, by the attention that was devoted to this problem at the Twenty-Fourth Congress of the CPSU. It can be resolved only on the basis of modern scientific and technical attainments: mathematical model building of processes and of the functioning of the economy and its individual elements; the use in planning and information processing of computers and other technical means integrated into automated control systems all the way up to the national level.

Mathematics and software in the broadest sense of the word — the whole gamut of descriptions, models, methods of algorithms, programs and languages, and systems for the organization and coding of information — are called upon to play an immense role in this work. It is with their help that actual production and economic activity and the management of this activity are transformed into an organized system of flows of information that is mathematically consistent and that also in-

---

*Vestnik Akademii nauk SSSR, 1972, no. 10. Academician Kantorovich's coauthors are doctors of science.

cludes the mechanical processing of these flows. Of particularly great importance are quantitative mathematical methods, chief among which are the methods used in solving extremal problems: optimal mathematical programming. It was not by chance that these methods, and especially linear programming, first appeared in the Soviet Union, since it is specifically under the conditions of a socialist economy that the problems of optimal utilization of resources for the good of society are broadly and systematically resolved. Linear optimization models proved to be an effective and universal means of economic model building and became widespread as the basic apparatus of optimal planning. They combine the flexibility and possibility of taking into account very subtle features of various economic processes in static and dynamic form with the existence of powerful computational methods that are universally suitable for calculations based on such models.

The methods of linear programming make it possible to elaborate not only an optimal plan but also a system of indicators (dual variables) coordinated with the optimal plan. This is especially important under the conditions obtaining in our economic system, in which all basic economic indicators must be planned and are planned together with production and distribution. Prices and their structure, depreciation allowances, capital charges, rent for the use of natural resources, normative effectiveness of capital investments, etc. — all these things have been strictly substantiated in specific mathematical models, which has also helped to clarify the nature of the economic categories themselves.

The use of mathematics in economics has influenced not only the apparatus of economic science but also mathematics itself. Such mathematical concepts as penalty functions and block programming methods have been induced by economic considerations. The theory of algorithmic languages is developing under the great influence of the need to resolve accounting and informational problems.

To a considerable degree the mathematical problems of economics have promoted the development of some important sections of mathematics: the general theory of systems of inequalities, the constructive function theory, convex analysis, graph theory, etc.

The practical use of linear programming has proven its high degree of effectiveness. At the same time, the basic difficulties and problems confronting researchers have been clearly articulated. Mention should first be made of computational difficulties. Even within the framework of linear programming, we are now dealing with problems of such dimensions that the capacities of existing and projected computers are inadequate for their solution by existing methods (later we shall return to the question of the manner and measure in which these difficulties are being resolved).

Further difficulties arise in model building. The linear model of the economy and the possibility of using it in a real economy are based on a whole series of assumptions — the existence of certain homogeneous ingredients (their unlimited divisibility is also implied), the linearity of production processes, the completeness and accessibility of information, its stability, and the existence of an indisputable optimality criterion. As we know, many of these assumptions do not always agree with reality. For example, in the variation of the intensity of production processes the hypothesis of linearity is correct only within certain limits beyond which there is an area of substantial nonlinearity resulting in discrete changes in output and material expenditures. Even in very detailed models there are numerous "extramodel factors" the actual influence of which is sometimes too significant to be left out of consideration. Certain factors cannot be included at all in determinate models.

Great difficulties arise in the selection of the optimality criterion, to which many indicators can lay claim in almost equal measure.

These difficulties are resolved by the creation of more precise mathematical models of optimization (nonlinear, dynamic, discrete, or partially discrete) as well as by the utilization of other concepts and tools in model building, such as the theory of games, the theory of controlled random processes, the statistical theory of collective behavior. In order to incorporate the economic "background" and specific features into optimization models, particular models are developed and studied in depth: models of price formation, of production location, of the effectiveness of capital investments, of equipment utilization, of supply and demand, of stock control, etc. The new models lead to the appearance of new objects of mathematical analysis and to new, still more complex computational problems.

The informational difficulties are connected with the fact that in optimization models the initial data on resources, needs, and the expenditure of material are considered to be known and precise. The problem of "filling models with information" is presently one of the most important problems. The danger of using inaccurate information is all the greater in view of the fact that in the optimization process the algorithm "seeks" not the most characteristic average values of parameters but rather the least likely "deviations." The advent of automated systems for the collection and processing of data offers hope for the expansion of the normative base for solving optimization problems.

Another aspect of the informational difficulties is the difficulty of obtaining forecast data. The problem of constructing long-term plans, and even planning for the immediate future, requires forecasts of needs, resources, relations, and technical potential. This leads to the necessity of constructing special forecasting models: models of technical progress and of demographic, economic-geographic, and social development. Further progress in methods of model building is, in turn, essential for the solution of these problems.

*   *   *

Let us examine in somewhat greater detail the ways of overcoming the enumerated difficulties. Models of linear programming and numerical methods for them continue to be a central part of the optimal planning apparatus. For this reason the expansion of the potential uses of these methods — the increase in the dimensions of the problems to be solved — continues to stand at the center of researchers' attention.

The dimension of the problem of linear programming is determined by two basic parameters — the number of constraints of a general type and the number of variables, which is usually greater than the number of constraints. Problems of large dimensions may have a large number of variables either with a comparatively small number of constraints (from several dozens to 200-300) or with a very high number (on the order of 500 or more). The problems of the first type are not truly complex. They are resolved by means of some method used in the economical array of information and in calculations with the use of external storage of computers or by programming the solution so as to include a unit for the generation of columns of the constraint matrix. The latter technique is effective in handling a number of practical problems: in the problem of optimal utilization of material, the unit in question yields layout cards; in truck dispatching it turns out assignments (trip tickets) for trucks, etc. It is only problems of the second type that are genuinely complex. The basic difficulty that arises in trying to solve them by means of the traditional finite methods of linear programming lies in the need to solve large linear systems that appear at every stage of the computational process. The method of sequential improvement of the plan by using as a multiplier the inverse of the matrix (while storing it in the computer's memory in the form of the product of an elementary type of matrices) is considered to be most effective in such cases. Soviet mathematicians have proposed economical schemes of so-called repetition that have significantly increased the effectiveness of the method. Further substantial progress can evidently be achieved only for individual classes of problems. In this connection we note that very effective methods were created for problems in which, with the exception of a small number of constraints that apply to the entire aggregate of variables, the remaining constraints apply solely to individual groups of variables. For example, in the branch planning problem we distinguish between general constraints and constraints that apply to control parameters for individual regions. Another case is the class of so-called transport problems which now includes dynamic, multidimensional, and other problems. The further development of these methods presupposes isolating new classes of problems and the broader application of individual methods while utilizing combinations of various approaches. We may cite the example of the algorithm for calculating the rational utilization of rolling mills. This algorithm was developed in Novosibirsk and has been used for distributing orders for a significant proportion of the output of rolled metal and pipe.

Particular attention has recently been devoted to the computational side of the calculations — to problems of stability and variant solutions and of the cumulation of errors in the process of calculation, etc., to providing information input and assuring the control of the computation process. Programming for the computerized solution of linear optimization problems includes both the creation of specialized packages of programs, each with its own language, with its own system for preparation and alteration of data, and the creation of module units that as control modules can be plugged into automated systems.

In addition to further improving finite methods, it is expedient to develop various types of infinite iterative processes, e.g., by using concepts of nonlinear programming and game methods. The attractiveness of the iterative methods is explainable in terms of (1) the opportunity they offer to use in the process of calculation information that is given in compact form, which frequently reduces the demands made on the core storage of the computer, and (2) the greater computational stability of infinite iterative algorithms as compared with finite ones. It is of interest to combine the ideas of finite and infinite methods in a single computational process.

The need arises more and more frequently today to use nonlinear models for the description of economic processes. The basic source of nonlinearity in these processes is the nonlinearity of the objective function. The concept of nonlinearity includes discontinuities that occur in connection with the discrete character of such decisions as construction of new capacities,

231

selection of one of several variants, as well as "smooth" non-linearities that reflect or simulate the heterogeneity of production expenditures. Nonlinear programming studies "smooth" nonlinear problems. One of the "customers" for nonlinear programming is . . . linear programming, since it has been found that it is sometimes expedient to inject nonlinear additives even into linear problems in order to improve the computational process. As a rule, however, the effective computational procedures of nonlinear programming are related to an increase in the volume of information that must be stored and processed at every step, as in many rapidly converging procedures the volume of information used grows as the square of the number of variables. Nevertheless, it is possible to devise procedures in which rational relationships are maintained between the rate of convergence and the volume of working information in the computer's memory.

In this connection intensive research has been conducted on iterative processes in which plans (permissible states) with monotonic growth of the objective function are constructed sequentially. These processes can be represented geometrically in the form of a sequence of transitions from one point in space to another, i.e., from one state to another. Here the transition from the current state to the next state consists of two stages: the choice of direction of movement and the choice of the length of the step. Important particular instances of this procedure — the conjugate gradient method, Newton's method, and similar techniques — as a rule assure quadratic or superlinear convergence. Their development and improvement appear to be very promising.

As we know, the delayed convergence of gradient methods in which the direction vector equals or approximates the gradient vector is explained by the elongation of surfaces of the level of the maximized function in one or several directions. Therefore transformations of space which transform "ridges" on the graph of the maximized function into "peaks" in individual steps of the iterative process, and which thus diminish this elongation, must accelerate the convergence. A number of interesting implementations of this idea have been proposed. The observance of the constraints that the plans being sought must satisfy in some cases presents a special problem the solution of which requires special methods. There are several fundamental ideas on which they are based.

One of the approaches is based on the concept of the permissible directions in which it is possible to move while remaining in the permissible area and increasing the objective function. The other approach uses the function of penalties: the problem with constraints is replaced by a problem without constraints by means of the introduction of a penalty component into the maximized function. The component in question contains a multiplier — the parameter of the penalty, the growth of which leads to an increase in the amount of penalty. With a sufficiently large penalty parameter the maximum point in the problem without constraints proves to be close to the optimum being sought in the problem with constraints. The third approach is based on the examination of the Lagrange function, which is used to construct a dual problem for determining the values of Lagrange multipliers and applying to it methods of the type described above. As a rule the latter method leads to algorithms that are too slow. The shortcoming of the penalty methods is that the growth of the parameter of the penalty causes a sharp deceleration of the convergence of methods being used to search for the unconditional extremum. Recently a method combing both of these approaches was proposed. Practical use of the method has demonstrated its effectiveness. Here the penalty component is added not to the objective function but rather to the Lagrange function, and after each iteration one recalculates not only the approximate value of the optimal point but also the value of Lagrange multipliers for the next iteration of the process. It has been found that with certain allowances the method converges with linear speed, and (what is especially important) the penalty parameter does not increase with each consecutive iteration.

The convex programming problem is closely associated with convex games. For this reason the methods of games are at the same time the methods of convex programming. The Braun method for matrix games (the simulated game method), which is now extended to a broad class of games involving many persons, has been significantly generalized.

The development of mathematical programming has given impetus to the development of general approaches to the analysis of extremal problems. The transition to infinite mathematical programming has made possible the creation of quite large general systems that encompass nonclassical extremal problems from the theory of approximations, the theory of moments, and other fields of mathematics, each of which was previously the subject of special research. This is probably the most vivid example of the development of age-old mathematical areas and of the ordering of them under the influence of the needs of economic science.

Owing to the difficulty of multiextremal problems, research on nonlinear programming is primarily directed toward the development of local methods that lead to optimal points only in the case of uniextremal problems. However, in order to solve practical problems, e.g., problems relating to the long-term planning of the development of branches and the siting of production, it is also essential to improve the methods of analyzing multiextremal problems. The ideas and methods of so-called discrete programming can offer a great deal in this respect.

Several different directions can be singled out in the solution of discrete extremal problems. First of all, we should mention work to isolate classes of problems in which the solution is obtained by means of more or less simple methods, e.g., the direct or slightly modified use of linear programming. Research associated with the deeper use of the methods of linear programming, and in particular with the so-called cutoff method, has for the most part been of a theoretical nature. While recognizing the importance and unquestionable timeliness of such work, we nonetheless note that practical needs for highly effective algorithms could not be properly satisfied in the given instance.

Considerably more has been done to improve computational methods based on the scanning of variants. Various ideas on reducing the scanning process have made possible the creation of stable work methods for a whole series of important practical cases: discrete problems of optimal layout (particularly in shipbuilding), the choice of optimal backup systems and error detection schemes in complex systems, and statistical problems entailed in the search for the formula of optimal approximation.

While research on the method of improved scanning has been under way in our country for a relatively long time, the computational significance of these methods was initially underestimated. Among discrete and nonconvex problems particular interest has been evoked by the problem of location of production and the development of branches, which is important for economic applications. A whole spectrum of methods has been tested in problems of this type: improved scanning, suboptimization, incomplete scanning, random search, and various combinations of computational ideas. Variants of this problem (for example, with additional constraints on total production, capital investments, etc.) have also been examined. Attention has now been attracted to various generalizations of the problem — to the dynamic problem of siting with the selection not only of the place or variants but also of timing of new construction, the multiproduct problem of siting considering limited possibilities of proposed construction sites, as well as interrelations of industries and their proximity to sources of raw materials.

This complex of computational problems can now rightfully be considered to be one of the most important for the application of the theory of mathematical programming, since calculations of the optimal industrial location have already encompassed more than half of industrial production — both with respect to the number of branches and with respect to capital investments.

Among other discrete problems that merit our unflagging attention, let us mention the choice of optimal schedules (especially in production and construction problems) and the determination of discrete production parameters. Work should be continued on the creation of combined methods, and in particular, we should use the combination of the cutoff method and the method of improved scanning with one another and with other approaches to optimization. We should learn to evaluate the volume of calculations required by the use of the methods of improved scanning for specific classes of problems. Dynamic programming is an important method for the solution of certain discrete type problems. Work on its development has also proceeded in several directions and on a wide scale. Thus in addition to the direct use of enumeration based on recurrent relationships, the method of recalculation of lists of states was developed. The Kiev mathematicians who proposed this method have used it to solve many important economic problems. Among them were such problems as choosing an optimal layout for a railroad bed, the optimal location of compressor stations on a gas pipeline, etc. The essence of the method lies in comparing intermediate states of the dynamic process in terms of "economic" indicators and then discarding those that are the "worst" in this sense. Thus at every step there is only a comparatively short list of "competitive states," and subsequent steps are made after the list has been recalculated.

This method can be regarded as a variety of the improved scanning method.

Dynamic programming is also important as a powerful apparatus for studying the probabilistic problems of optimization when in the choice of controls it is necessary to consider that the future course of a process depends on random factors. The plan for optimal prevention and for backups for unreliable and worn equipment, flaw detection, inventory control — all these problems can be incorporated in the scheme of dynamic programming processes (semi-Markovian decision-making processes).

Matters are much more complicated with respect to probabilistic analogues of linear programming problems that are studied in stochastic programming. The practical need for such schemes is determined by the fact that observations and forecasts are sources of our knowledge about the parameters of the process under investigation. In both cases the information contains errors that in the first approximation may be described by means of probabilistic models. Here the difficulties arise even in the model preparation stage, since it is necessary to describe rather sophisticated planning and management situations in the face of risk and uncertainty. In particular, there are many objective functions differing in sense, the selection of which is determined depending on the problem at hand: the mathematical expectation of income or some other linear function from the plan, the dispersion of this function, the probability that it will exceed a certain level, etc. A different sense is also invested by stochastic programming in the concept of constraints (on the average, according to probability, etc.) and in the concept of the plan. The plan can be defined as a determinate or as a random vector. In the latter instance the structural dependence of the plan on certain initial random values is usually selected beforehand (most often from the physical sense of the problem and sometimes on the basis of considering the convenience of its subsequent analysis). Even now it is possible to speak of some successes in the theory and use of stochastic programming. However, only the first steps have been taken toward actual application to questions pertaining to the probability assessment and preliminary processing of data prepared for optimization calculations. This problem will require the development of uniform methods incorporating optimization and statistical approaches. In particular, it would be expedient to develop methods based on multiple-step and directed accumulation of statistical information for the subsequent refinement of decisions.

Aggregation — the amalgamation of branches and products into larger groups — is another important problem entailed in preparing data for economic models. Recently there has been a certain amount of progress in this area: approaches have been defined more precisely, and iterative methods have been constructed for the calculation of aggregated models. This important research should be continued and developed.

* * *

The improvement of the data base, the elaboration of forecasts, analysis of the results applying mathematical methods, the ascertainment of new areas for their application — all these require painstaking work on the creation and study of mathematical models of the economy.

The elaboration of the optimality criterion or the objective function for the global model of optimal planning is a difficult problem in economic optimization model building. The criterion of maximization of consumption for quite an extended period of time is substantiated in the works of some scholars. It should be noted, however, that in certain cases the basic conclusions of the theory of optimal planning do not depend on a specific type of objective function.

The general model of optimal planning is convenient for theoretical purposes. On the other hand, the practical application of the model in its full form is difficult. By simplifying the general model the Institute of Mathematics of the Siberian Branch of the USSR Academy of Sciences obtained a large aggregated model for long-term optimal planning. Slightly different types of applied models have been proposed by the Central Economic-Mathematical Institute, by the Institute of Economics and Organization of Industrial Production of the Siberian Branch of the USSR Academy of Sciences, and by the Economics Institute of USSR Gosplan.

In theory, large-scale aggregated models for the purposes of consolidated national economic planning have already been quite well elaborated. It is now important to test them and to incorporate them in the system of national economic planning. Of the mathematical problems that arise in the process, let us mention the elaboration of algorithms for calculating the influence of the fluctuations of initial data on various economic parameters.

As regards the theoretical study of optimization models of the economy, the interest of researchers has shifted to dynamic economic models that hold great importance for the development of the methodology of compiling long-term development plans.

Mathematically speaking, the dynamic model of optimal long-

term planning is a problem of finding a trajectory that origi-
nates in a given state and results in a maximum value of the
objective function related to the entire set of trajectories. In
this connection one may consider trajectories of both finite and
infinite length.

Among the optimal trajectories there are special trajectories
in which the basic proportions remain invariant. Such trajectories have come to be called turnpikes. Turnpikes possess
two remarkable properties. First, they are used to realize the
maximum possible continuous growth rate of the economy that
can be sustained for any length of time. Second, the arbitrary
optimal trajectory, no matter what its original state, over time
comes closer and closer to the turnpike — optimal trajectories
gravitate toward the turnpike (so-called turnpike theorems).
Turnpike theorems do not eliminate the problem of the optimality criterion in dynamic optimization models of the economy,
but nonetheless, they reduce the number of reasonable optimality criteria. In long-term planning they are used in estimating
the influence of the postplan period. This area has its unresolved problems: for example, the problem of fully assessing
technical progress in its interrelationship with the state of the
economy.

Soviet scientific literature devotes less attention to the study
of models of economic equilibrium than to optimization models.
This is natural, since the models of equilibrium are of a less
constructive nature and aim only indirectly at the improvement
of the planning and management of the economy. Unlike models
of optimal planning, they do not presuppose the existence of a
single goal for various participants in the economic process.
The result of interaction may lead to certain stable states that
are called states of equilibrium. In terms of the subject matter
of mathematics, the theory of economic equilibrium is close to
the theory of games involving many persons.

It is important that we establish conditions under which the
states of equilibrium prove to be optimal, i.e., result in the
maximum value of a certain global objective function. The ascertainment of these conditions holds decisive importance for

economic theory, especially for making possible the decentralization of management. The solution of this problem is also important for the construction of an incentive system for individual parts of the economy that is coordinated with the general objectives of society.

Of interest are mixed models in which the model of economic equilibrium is only a part, since in the actual economy processes take place that are associated both with equilibrium and with global optimization.

The elaboration of numerical methods for finding states of equilibrium still requires a great deal of effort. Of particular interest are methods the computational layout of which models a certain process of movement toward a state of equilibrium.

The models that have been examined thus far have pertained to the entire economy as a whole. However, they are called models of the economy only conditionally, since as a rule they reflect only one aspect of interrelations between various parts of the national economy: the global balance aspect. To fit these models even better to the practice of planning is especially important in connection with their use in numerous automatic control systems; it requires our unflagging attention to schemes describing in greater detail the functioning of economic systems.

Individual studies of this nature have already been conducted and are presently continuing. The intention is to solve problems of inventory control and of PERT and calendar planning and to compile models for the optimal, timely replacement of equipment (from the standpoint of its reliability and cost indicators). Research began on models for supplying products with uneven seasonal consumption. In such cases models of transportation and inventory control are combined. However, their fitting into the models of general economic systems has not yet been sufficiently investigated.

Another important parameter in the economic system, which can be selected only with due regard to the functioning of the system, is the structure of its management. The determination of the structure of subordination, of levels of responsibility, and of information flows is only part of the problems that arise

in the process. Problems pertaining to the construction of a
system of national economic models (in particular, the question
of the operation of a system of models) must also be resolved
from the standpoint of models of functioning. The problem of
combining branch and territorial planning is also extremely
urgent.

* * *

The foregoing discussion makes clear how numerous, complex, and diverse are problems associated with the creation of
software required for the modern scientific management of the
economy. It is clear that economics makes no fewer demands
on mathematics than does physics or mechanics. The immense
practical significance that the level of development and mastery of software holds for improving the planning of the management of the national economy is evident. At the same time,
the scientific achievements in this area, the scale of work, and
the personnel fall short of the size and urgency of the problem.
This is why it is so urgent to increase the breadth and depth
and to conduct further intensive research on mathematical methods of economics. The vital necessity of this research and the
strict deadlines that are imposed by practice require special
organizational measures. Among these measures we should
mention first the transfer of a considerable number of mathematicians to work in the new area. This would be furthered by
the creation of corresponding laboratories and departments in
mathematical scientific institutes, higher educational institutions, as well as economic planning research institutes and
computer centers. In the latter instance in our opinion, we
should establish, in addition to production groups, special research groups that are relieved of routine operational work.
The departments of mathematics should introduce corresponding specialization for students and should organize retraining
for mathematicians. As experience has shown, no small part
could be played by brief "winter" and "summer" schools. It is
essential to organize the training of mathematician-statisticians

241

and to work out plans for training technical personnel specialized in mathematics. We should also elaborate a program for creating special textbooks and monographs and possibly should create a special journal on these new fields of mathematics.

18

Optimal Utilization of
Rolling and Pipe Mills*

L. V. KANTOROVICH

On the basis of many years of experience in the management
of the socialist economy of the USSR and in scientific planning
of the economy, particularly the experience of the work of supply agencies, it is possible to draw general conclusions concerning ways and means of realizing the advantages that are inherent in the socialist economic system.

This experience is attracting particular attention during the
days of the Lenin jubilee, since the behests of Vladimir Il'ich
Lenin on methods of socialist construction have been implemented in the more than half-century of development.

Analysis of the solution of problems that have arisen in the
last few years is of special interest. On the one hand, these
problems are associated with the implementation of the economic reform directed toward the further improvement of the
planning and management of the national economy, and on the
other hand, they are connected with the broad introduction of
modern scientific methods and cybernetic machines in the management of production.

In our view it is specifically these aspects that make it possible to elicit to the maximum — in addition to the well-known

---

*Material'no-tekhnicheskoe snabzhenie, 1970, no. 4.

and generally acknowledged socioeconomic advantages — the organizational and economic advantages and the productive effectiveness of the socialist system compared with the capitalist system.

While in its day mechanized production made it possible to reveal the economic advantages of capitalism compared with feudalism, the transformation of science into a direct productive force and the use of cybernetic machines that have augmented man's thinking activity many times over can be fully realized only under socialism. The possibilities of using the advantages of socialism are especially obvious in this respect. They can be traced, in particular, in the work of USSR Gossnab. The Western world does not have a firm of such size and potential, that is so powerful and complex in terms of its tasks and organization.

The present report briefly depicts a specific job that was begun at Soiuzglavmetall at the initiative of the Institute of Mathematics of the Siberian Branch of the USSR Academy of Sciences, in which we directly participated. We are talking about the optimization of the utilization of rolling mills and the issuing of schedule-orders for metal by means of mathematical methods and computers.

The possibility for the rational distribution of production programs between rolling mills was noted as far back as 1940, immediately after the principles of the theory of linear programming appeared in the USSR. Since that time a number of attempts have been made, both in our country and elsewhere, to solve individual problems in the utilization of rolling mills. However, systematic work on the solution of this problem in full measure began six or seven years ago. The work is now nearing completion. In addition to the Institute of Mathematics of the Siberian Branch of the USSR Academy of Sciences, the following took part in the work: the Institute of Cybernetics of the Ukrainian Academy of Sciences, the All-Union Scientific Research Institute of Pipe [VNITI], the All-Union Scientific Research Institute for the Organization of Production and Labor in Ferrous Metallurgy [VNIIOchermet] of the USSR Ministry of

Ferrous Metallurgy, computer technology departments of Soiuzglavmetall and USSR Gossnab, and the Computer Center of USSR Gossnab.

What is the essence of the work that has been done? What was previous practice like, and what constitutes the newly developed and implemented system?

Soiuzglavmetall put the placement of orders for metal on a centralized basis. In accordance with the funds allocated by planning organs in a large aggregated nomenclature and coordinated with the production plans of metallurgical enterprises (for specific mills), on the basis of the customer's concrete specifications this main administration issued them warrants to receive the required products from certain enterprises. At the same time, it was necessary to take into account the specialization and technical potentials of metallurgical plants and individual mills, production constraints regulated by special protocols, shipping demands, etc. The enormous volume of the work (for example, for small grades it was necessary to prepare tens of thousands of warrants on the basis of a nomenclature that numbered several thousand designations and profiles and sizes) and the need to perform this work in the shortest possible time created great difficulties. In the process of the work there also arose other complications, since the planning balances were kept in large aggregated form, some labor-intensive profiles and dimensions of metal proved to be in short supply, and it was necessary to limit the demand for them and in some cases to replace them with others. In the process of issuing schedule-orders there was no search for the optimal solution — optimization as a whole. In this process there was an attempt to observe the principle of territorial proximity between the manufacturing plant and the customer and in this regard to get some reduction in average shipping distance.

The use of mathematical methods to solve this problem revealed that the productivity of mills differed for different nomenclatures, and thus the placement of orders affects their actual productivity. However, one cannot be guided by this factor alone: It is also important to consider the distance over

245

which the metal is shipped. For this reason the problem was posed of optimizing the placement of orders, with due regard to the requirements of customers and the constraints imposed by suppliers, so that the productivity of the mills would be maximal while the volume of transport costs would not exceed a given level (another combination of these requirements is also possible: for example, the minimization of the sum of production and transport costs).

This is a linear programming problem. If it were viewed as a problem of a general type, it would have gigantic dimensions — a matrix on the order of $3,000,000 \times 30,000$ (variable constraints). Evidently, problems of such a volume have never before been dealt with. However, the special character of the indicated problem has facilitated its description and solution.

Mathematicians had to elaborate special methods and algorithms that were first programmed and tested on the M-20 computer at the Institute of Mathematics of the Siberian Branch of the USSR Academy of Sciences; then the system of programs was reworked for the Minsk-22 computer, which metallurgical institutes and the Computer Center of USSR Gossnab have at their disposal.

Metallurgists had to gather and analyze information on the productivity of hundreds of mills for thousands of types of assortments. It was necessary to elaborate a coordinated system for representing and processing information (orders, constraints).

The transition from manual work to computerized optimization calculations required a radical change in the system of issuing schedule-orders, some change in the thinking of personnel — overcoming a psychological barrier, and the operational solution of a number of problems arising in the course of the system's realization. Thus, for example, it proved necessary to significantly alter the production plans of metallurgical enterprises compared with original plans.

The optimal utilization of rolling mills is based on the existing centralized system of metal supply. Without such a centralized planning system it would be impossible to attain 100 percent utilization of the capacity of rolling mills. The measures that have been implemented in this direction should be viewed as an improvement in the aforementioned system. However, this improvement is basic and permits the much more complete and effective utilization of its advantages and possibilities.

The goal of the reorganization lay not only and not so much in transferring the bulk of the work associated with schedule-orders to the machine but in significantly improving their quality. Experience has shown that this goal is attainable. Schedule-orders calculated by optimal methods on computers proved to be more effective.

In the case of gas pipelines alone, for which the system was first put to work, the rational placement of orders with enterprises — with due regard to the volume of orders on the one hand and production capacities on the other — revealed the possibility for the additional production of approximately 60,000 tons of products with the same production capacities. At the same time, part of the reserve was expended on fully eliminating the shortage of labor-intensive types of products. Transport costs were reduced by approximately 15 percent. A similar result was obtained for a number of other types of rolled metal to which the optimal system of schedule-orders was also applied (welded pipe, medium size, etc.).

In the second half of 1969 the optimal system of schedule-orders was applied to approximately 20 million tons of rolled metal; the system is slated to be applied to other types as well (small size, sheet metal, etc.). The application of this method to all types of rolled metal will be the equivalent of producing additional hundreds of thousands and even millions of tons of rolled metal. But if we consider the fact that enterprises are now issued orders in precise accordance with their production capacities and that they are interested in their maximum disclosure and expansion, the attained increase in capacities as a result of the new system may prove to be the equivalent of an annual 4-6 percent increase in the capacities of rolling mills. The elimination of the shortage of a number of rolled profiles is of great economic importance.

Work is now continuing on the improvement and expansion of

the application of these methods. They can be applied, for example, in the production of automobile tires, paper, etc. Despite the completely different qualitative nature and conditions of the production of these types of products, the given methods and the system for their application may in large measure be used for them as well.

Along with the optimal distribution of orders, much interest attaches to the use of the simultaneous solution of another problem: estimates of the labor intensiveness of production and of production capacities. They can be used as the basis for the objective calculation of the optimal system of prices. The establishment of scientifically substantiated prices on metal products will make it possible to achieve a situation in which all rationally placed orders will prove to be equally advantageous for the enterprises. This will still further facilitate the optimal utilization of mills and schedule-orders.

To date these operations have been performed on computers of average capacity. USSR Gossnab now has the powerful "system 4-50" computer. The use of this system will make it possible to combine work on the optimal utilization of mills with the processing of all data on the course of fulfillment of orders, which will permit the still more efficient control of their fulfillment.

Work is now also being performed on the coordination of centralized management of supply with the activity of territorial agencies. Evidently, in the future it will not be expedient to place all orders directly with the center but rather to transfer some of them to territorial administrations. However, the proposals to directly empower territorial administrations with discretion concerning the utilization of the corresponding production capacities of suppliers for the realization of orders received from the customers in their regions are hardly acceptable. Such a division of the problem of placement of orders violates the optimality of the solution and reduces the effectiveness of utilization of production capacities. There is greater advantage in the further iteration of the problem — the unification of questions pertaining to the siting of the production of

various types of rolled metal and to the inclusion of the planning
of orders for semifabricated goods, raw materials, etc.

The problem of taking account of demands for the stability of
economic relations is also an urgent one where optimal distribution is concerned. While their accounting is possible in
principle, it has not yet been computerized. It should be pointed
out, however, that such stability is not only not always possible
(for example, owing to the introduction of new production capacities) but also is not always necessary. If we are discussing
technical and production relations for supplies that are specific
for a given type of production, the stability of relations is important. But if we are discussing standard supplies with the
standards being precisely observed, it will probably be more
correct to distribute them either directly to the customer or
through territorial administrations.

The experiment conducted to achieve the optimal utilization
of rolling and pipe mills and the associated scientific analysis
of problems pertaining to the placement of orders for rolled
metal make it possible to draw certain general conclusions
concerning the further improvement of supply.

First, it appears evident that USSR Gossnab cannot confine
itself solely to the distribution of finished supplies. The placement of orders enables it to exert an active influence on production and the volume and specialization of production and hence
to determine more precisely what and how much should be distributed. For this reason USSR Gossnab should study production conditions more thoroughly and should maintain closer relations with production ministries.

The second point that has been confirmed anew by the practice of optimal utilization is the need to correct production
plans and to make operational and effective changes in them.
For example, when it was discovered that an additional 60,000
tons of gas pipe could be produced, the next problem was to
ascertain additional customers and to determine whose needs
should be satisfied. There also arose the problem of finding
additional raw materials for the production of pipe.

Thus the active forms of organizing the circulation of the

means of production require that supply agencies have certain rights and that they participate independently or jointly with planning agencies in the operational correction of production plans.

Operational regulation would be facilitated if USSR Gossnab, in addition to supplying customers, were to create certain own reserves (undistributed reserves), especially of standard types of products. Such centralized undistributed reserves would be much more effective than reserves scattered among the various enterprises; they would permit more effective assistance to the enterprises.

The necessity for the participation of supply agencies in operational distribution is also justified by the fact that when they have continuous contact with the customer, these agencies know for which products the actual need has declined or increased. By having available undistributed reserves of supplies, they can assure additional output, since at enterprises in many branches output is limited not so much by production capacities as by the shortage of supplies. Additional output in excess of the plan, particularly at the initiative of the enterprises themselves (for example, the production of a series in place of one item made to order or new types of products), should receive wide support.

Evidently, in the process of development of the economic reform, we must expect a general transition to a more flexible system of planning of production so that the enterprise's plan is determined on the basis of forecasts of its potential and the need for the product and is amended in the process of fulfilling the target. This would promote the more complete disclosure of reserves. In turn, such an increase in the flexibility of production planning will also require greater flexibility in the system of supply. Under these conditions the supply plan must be based not only on planned requisitions (to the extent that they have been specified) but also on the forecast estimate of additional requisitions that still remain to be specified. Provision should be made for a flexible operational system of the regulation of supply that envisages backup resources, a system of priorities in requisitions, economic means for influencing the volume and time of deliveries, etc. It is expedient, for example, that 80-90 percent of the orders be filed with precise specifications and that in the case of the remaining orders the right be granted to correct the nomenclature, and perhaps also the volume, of deliveries in the course of the year. However, considering the interest of producers and planning and supply agencies in receiving early information concerning orders, it is expedient to pay 10-15 percent more for such additional requisitions, so that the customer would use them only if it were necessary and advantageous to him. Here an important role must be played by a flexible price system that takes real expenditures and the scarcity of products into account.

Thus the further development of planned supply under the conditions of the economic reform should be characterized, on the one hand, by the increasingly broad use of scientific methods and the technical means of cybernetics and, on the other hand, by the lessening of rigidity in the planning of supply, by greater efficiency in the satisfaction of requisitions, and by the replacement of constraints by economic levers.

The high degree of flexibility and efficiency of centralized supply permit it to acquire many of the features and advantages of wholesale trade. The question as to expedient forms of combining centralized supply and wholesale trade requires further elaboration with due regard to the specific conditions that assure the greatest national economic effect.

An important role here belongs to the union of science and practice. Modern science has a great reserve that can find practical application in the national economy (e.g., in the organization and planning of material-technical supply). At the same time, in the process of its intercourse with practice, science itself receives new and interesting problems and the possibility of generalizing practical experience and of discovering new ways to further increase the long-run effectiveness of socialist production.

About the Editor

Leon Smolinski is a professor of economics at Boston College and an associate of the Russian Research Center at Harvard University. He also taught at the Massachusetts Institute of Technology and at the University of Michigan, Ann Arbor. He is a member of the editorial board of Problems of Economics.